Life Cycle Assessment in the Built Environment

To a friend and an inspiration

1969–2008
Associate Professor Graham Treloar

Life Cycle Assessment in the Built Environment

Dr Robert H. Crawford

Routledge
Taylor & Francis Group

LONDON AND NEW YORK

First published 2011 by Spon Press

This edition published 2013 by Routledge
2 Park Square, Milton Park, Abingdon, Oxon OX14 4RN

52 Vanderbilt Avenue, New York, NY 10017

First issued in paperback 2020

Routledge is an imprint of the Taylor & Francis Group, an informa business

© 2011 Robert H. Crawford

Typeset in Frutiger by Saxon Graphics Ltd, Derby

British Library Cataloguing in Publication Data
A catalogue record for this book is available from the British Library

Library of Congress Cataloging-in-Publication Data
Crawford, Robert, 1978-

Life cycle assessment in the built environment / Robert Crawford.

p. cm.

1. Building materials--Service life. 2. Buildings--Environmental aspects. 3. Building materials--Recycling. 4. Product life cycle. I. Title.

TA409.2.C73 2011

624.1'8--dc22

2010032823

ISBN 13: 978-0-367-57698-1 (pbk)
ISBN 13: 978-0-415-55795-5 (hbk)

Contents

Figures

Tables

Foreword

The benefits of modern urban settlements are self-evident. Towns and cities provide housing, employment, education, healthcare, public transport and many other services. However, the construction and operation of buildings and infrastructure consume significant resources and produce large quantities of waste. We are now at a stage where we are questioning whether the development path taken in the past is suitable for the future and this is the starting point for this book.

The case for analyzing buildings over their whole life is logically made and this is tied to environmental impacts including the consideration of raw materials, energy, water, emissions, wastes and other indicators as an introduction to life cycle assessment (LCA). The principles, framework and guidelines for LCA, which are now part of international standards, are described in a clear and easily understandable manner.

What follows is the truly innovative part of this book. The combination of embodied energy theory and LCA are integrated into a streamlined environmental assessment approach. Furthermore, input-output-based hybrid analysis is described and used to show how a much more comprehensive assessment of environmental impacts in the built environment is achievable compared with current methods.

This book adds clarity to the uncertainties in life cycle assessment and provides a clear methodology to ensure consistent analyses. Hence, the innovative contribution is to overcome difficulties with system boundary definition and the life cycle inventory data. The text draws upon the work of the late Associate Professor Graham Treloar especially with respect to his innovative pathway extractor tool for input-output analysis. This research has been developed further by the author of this book and the contribution that this body of work has made to LCA is clearly described. *Life Cycle Assessment in the Built Environment* is essential reading for students, researchers and practitioners in the field. Chapter 5 will be especially appreciated as it provides seven case studies which show the exact methods used for conducting streamlined life cycle assessments. Finally, a series of initiatives are proposed for creating more sustainable built environments in the future. This will involve some very challenging choices and this book will play a significant part in that process.

Stephen Pullen
University of South Australia, Adelaide
July 2010

Preface

The need for a detailed understanding of the impact that human activity has on the environment, in order to make more sustainable choices, is becoming more important as an awareness of the broad environmental issues that society is currently facing grows. With the construction and operation of the built environment (our buildings and infrastructure) responsible for a significant proportion of these impacts, those involved across all stages of the built environment life cycle must be armed with the tools and knowledge needed to significantly improve its environmental performance. This is essential if we are to avert potentially catastrophic environmental damage and the threats to our very survival that this may bring.

This book came about by the need to address some of the limitations of existing approaches for assessing the environmental performance of the built environment. It brings together much of my own work in the development and application of hybrid life cycle assessment over the past ten years. More broadly, there has been considerable work done in this area, particularly over the last two decades, with little more significant than that of the late Associate Professor Graham Treloar who, as a passionate advocate for sustainable construction, developed an innovative approach for comprehensively quantifying the environmental impacts caused by, amongst other human activities, the construction of the built environment. This book is an attempt to continue and extend this work as well as make it more accessible to the decision makers in the global construction industry, so that we may collectively work towards a more sustainable approach to building, operating and managing the built environment.

The book begins by describing some of the key environmental issues that we are currently facing and how the built environment is responsible for a significant proportion of the impacts that humans are having on the environment. Chapter 2 addresses some of the approaches and strategies that must be employed in order to slow down and ultimately reverse the environmental damage associated with the design, construction and on-going use and management of the built environment and justifies the importance and need for *environmental assessment* as an integral part of this process. The stages involved in conducting a life cycle assessment are described in detail in Chapter 3. A more specific description of the application of life cycle assessment within the built environment is provided in Chapter 4, using a single detached residential building as a case study to demonstrate the potential benefits and limitations of using life cycle assessment to inform better environmental outcomes within the design and management of the built environment. Chapter 5 presents some further life cycle assessment studies on a range of built environment case studies, ranging from whole buildings, building components, transport infrastructure and renewable energy technologies. The final chapter discusses the role that some of the key stakeholders in the built environment must play in helping to improve the environmental performance of the built environment. It also highlights some of the current barriers for achieving this, including some of the limitations of current environmental assessment approaches and how, in the future, these may be minimized. Finally, the importance of the integration of environmental assessment into everyday industry practice, and how this may help the professionals involved in the design, construction and on-going management of the built environment realize the full potential of environmental improvement opportunities, is discussed.

Acknowledgements

There are numerous people that have provided advice and expert knowledge that has helped to direct this book and I am truly grateful for their time. No one person has had more of an influence on the development of this book than a long-term friend and colleague in the late Associate Professor Graham Treloar. As an international leader in the methodological development of hybrid life cycle inventory approaches and a passionate advocate for sustainable construction, he has provided the inspiration for this book and my own dedication towards improving the environmental performance of the built environment. I will be forever grateful for the short time we shared and for the opportunity to carry on his legacy.

The following organizations provided information for the case studies presented in Chapters 4 and 5 and I am grateful for their kind support and their permission to use this information: Metricon Pty Ltd for the house plan and a bill of materials for the case study house used in Chapter 4, and the Victorian Department of Infrastructure (Australia), in particular Mr David Hill and Ms Kate Murphy, for permission to reproduce the findings from a study of the environmental performance of alternate railway sleeper types. Thank you also to Professor Manfred Lenzen (University of Sydney, Australia) for permission to use the Australian input-output models developed by him for 1996–7. I am also extremely grateful to Dr Stephen Pullen (University of South Australia) for his friendship and encouragement and also for his invaluable input on earlier versions of the text that has helped to direct this book. And most importantly, thank you to my family for their continuing and unwavering support, before, during and since undertaking this project.

Abbreviations

A$	Australian dollar
ABS	Australian Bureau of Statistics
BiPV	building integrated photovoltaics
CAD	computer-aided design
CFCs	chlorofluorocarbons
CH_4	methane
CO_2	carbon dioxide
DER	direct energy requirement
EE	embodied energy
FC	fibre cement
FG	fibreglass
GDP	gross domestic product
GHG	greenhouse gas
GJ	gigajoule (10^9 joules)
GWP	global warming potential
HFCs	hydrofluorocarbons
HVAC	heating, ventilation and air conditioning
HWS	hot water system
I-O	input-output
IPCC	Intergovernmental Panel on Climate Change
ISO	International Organization for Standardization
kL	kilolitre (10^3 litres)
km	kilometre (10^3 metres)
kW	kilowatt (10^3 watts)
LCA	life cycle assessment
LCI	life cycle inventory
LCIA	life cycle impact assessment
MDF	medium density fibreboard
MJ	megajoule (10^6 joules)
MW	megawatt (10^6 watts)
Mt	megatonne (10^9 kilograms)
N_2O	nitrous oxide
PET	polyethylene terephthalate
PFCs	perfluorocarbons

PV	photovoltaic
PVC	polyvinyl chloride
SETAC	Society of Environmental Toxicology and Chemistry
TER	total energy requirement
VOCs	volatile organic compounds
W	watt

1 Global environmental issues and the built environment

'If we do not change our direction, we are likely to end up where we are headed.'

Chinese Proverb

As members of the human race we are facing one of the, if not *the*, greatest challenges that we have ever had to face. Evidence now exists to support the fact that human activity is having a dramatic and potentially irreversible effect on the ecological systems that support our very existence. The current way of living in many countries around the world is responsible for immense adverse impacts on the natural environment. These impacts continue to increase despite a growing awareness and recent attempts at their mitigation. An increasing demand for the world's resources (particularly from developing countries), partially fuelled by a rapidly increasing population, and our technological- and consumerism-based societies are some of the main reasons why these impacts are a growing concern.

Almost all human activity results in impacts on the natural environment. Many of these impacts have the potential to result in long-term environmental degradation, with associated consequences for human health and, potentially, our very survival. These impacts stem from current human activities and the processes that we rely on to meet our current needs and standards of living and include climate change, the depletion of natural resources, the generation of waste and the release of pollutants into the environment.

Addressing the potential impacts of climate change is currently one of the most pressing issues, with evidence suggesting that these impacts may have a critical effect on both natural and human systems (IPCC 2007a). Humans, like many other Earth-based creatures, are highly susceptible to minor variations in climate. Rising sea levels caused by melting ice caps and the expansion of sea water through carbon dioxide absorption has the potential to wipe out thousands of people and habitats, particularly in low-lying regions of the world. Increased drought, rainfall and storm frequency has significant potential to displace millions of people and restrict access to some of the necessities for human survival, such as food and water. An increase in extreme temperatures, already being evidenced in many parts of the world in recent years, has the potential to contribute to an increase in human mortality.

Our reliance on fossil-fuel-based energy systems in particular has been shown to be one of the driving forces behind the rapidly increasing levels of certain greenhouse gases in the atmosphere, leading to global warming and the many and varied consequences of a changing climate. Also, our current rate of consumption of many of the Earth's resources and the practices we use to extract or remove them from the ground are rarely sustainable. Many of these resources cannot be naturally replenished at the same rate at which they are being consumed and traditional extraction and removal practices are also contributing to considerable environmental degradation.

The built environment alone is responsible for a significant proportion of these impacts, which result from not only the operation or use of buildings and other infrastructure but also from the extraction of raw materials right through to their eventual disposal. In order to alleviate many of the environmental issues that humans are currently facing, the built environment must be central to any environmental improvement efforts.

This chapter provides the background to some of the current environmental issues and their significance and establishes the context for the use of life cycle assessment (LCA) in the built environment. It will introduce the various systems and products that form the built environment (with a particular focus on buildings), the life cycle stages of the built environment and the most significant environmental impacts attributable to its construction and on-going operation.

1.1 Global warming and climate change

The current global warming trend and the potential catastrophic impacts that this may have on the Earth and its inhabitants is the most significant and challenging threat to its existence that the modern human race has ever had to face. Human-induced global warming is the result of increased concentrations of human-produced greenhouse gases in the atmosphere. Greenhouse gases absorb and emit solar radiation and the rate at which this occurs has an influence on global temperatures. As these gases build, the ability for long-wave heat to escape from the Earth is reduced, thus increasing the temperature of the air within our atmosphere (Figure 1.1).

Whilst certain levels of greenhouse gases are essential for maintaining air temperatures necessary for life on Earth, much of this life as well as many of the Earth's ecosystems are highly susceptible to even minor temperature variations. The most common greenhouse gases in the Earth's atmosphere are water vapour (H_2O), carbon dioxide (CO_2), methane (CH_4), nitrous oxide (N_2O), ozone (O_3) and chlorofluorocarbons (CFCs). Other than water vapour, which is not significantly affected by human activity, concentrations of carbon dioxide in the atmosphere contribute the greatest to global warming. While the global warming potential (GWP), or ability of a gas to contribute to global warming, of carbon dioxide is much less than for other greenhouse gases such

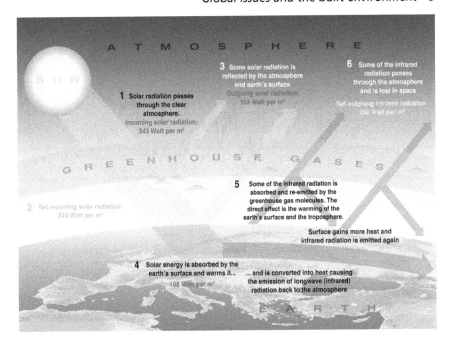

Source: UNEP/GRID-Arendal 2002.[1]

Figure 1.1 The process of global warming

as methane (with a GWP 25 times that of CO_2) and nitrous oxide (with a GWP 298 times that of CO_2) (IPCC 2007b: 212), its abundance means that it is the most significant human-influenced contributor to global warming.

Many of these greenhouse gases exist naturally in the atmosphere and are produced from the Earth's ecosystems. The atmospheric concentrations of greenhouse gases from the Earth's natural processes had remained relatively constant for at least the last 10,000 years prior to the industrial era (IPCC 2007b). Since these pre-industrial levels, concentrations of carbon dioxide, in particular, have increased by at least 38 per cent, from 280 parts per million (ppm) (IPCC 2007b) to the 2009 global annual average of 387 ppm (Tans 2010), as seen in Figure 1.2.

The greenhouse gases contributed through human activity are known as anthropogenic emissions. Human activities that contribute the greatest include the burning of fossil fuels, clearing of land and forests, certain farming practices (such as the use of fertilizers), industrial processes and waste decomposition. As Figure 1.3 shows, fossil fuel use was the single largest contributor to global anthropogenic greenhouse gas emissions in 2004.

Energy is a basic requirement of current human civilizations, a central necessity for many industrial processes and transportation systems, and crucial for the provision of amenities such as heating and lighting in our buildings. The emission of greenhouse gases at the power stations used to produce this energy accounts for 25.9 per cent of global greenhouse gas emissions. Industry

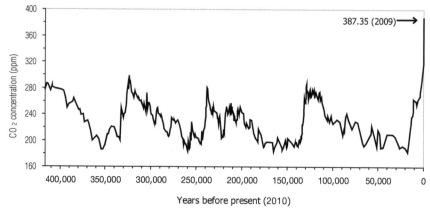

Source: Based on data from Petit *et al.* 1999 and Tans 2010.

Figure 1.2 Global trend in atmospheric carbon dioxide (CO_2) concentrations

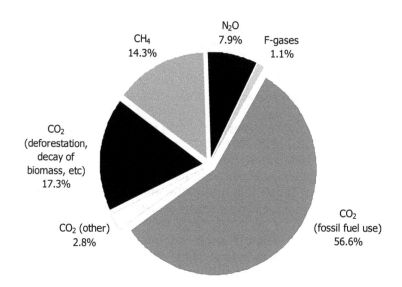

Source: IPCC 2007c: 103.

Note: F-gases are Fluorinated gases (such as HFCs, PFCs and CFCs).

Figure 1.3 Global anthropogenic greenhouse gas emissions, 2004

is responsible for the next largest proportion of emissions at 19.4 per cent. Figure 1.4 shows the total global greenhouse gas emissions by sector for 2004.

The Intergovernmental Panel on Climate Change (IPCC) has suggested that global temperature increases, very likely to be as a result of an escalation in anthropogenic greenhouse gases in the atmosphere, have been occurring at a rate of 0.13°C per decade since the middle of the twentieth century (IPCC 2007b). While this rate of average annual temperature increase may not be

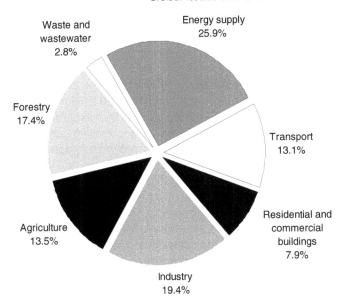

Source: Adapted from IPCC 2007c.

Figure 1.4 Global greenhouse gas emissions by sector, 2004

that noticeable to humans, this level of temperature variation can lead to significant changes in the Earth's physiological and biological processes. There is also a complex interrelationship between air temperature and other climatic processes. For example, it is suggested that the increasing concentration of greenhouse gases in the atmosphere and the resultant increases in global temperatures are also leading to a rise in sea levels, severe flooding and droughts, an increased frequency and intensity of extreme weather events and an increased incidence of disease (IPCC 2007a). These changes have significant potential to result in catastrophic consequences for millions of people living in the most susceptible regions of the world, particularly those in low-lying and tropical areas.

Health consequences for millions of elderly and young people could be immense, with recent extreme temperature events like those in Melbourne, Australia in January 2009 responsible for an up to 62 per cent increase in mortalities during these periods (Cooper 2009). The World Health Organization predicts that as many as 154,000 deaths in the year 2000 were attributable to climate change and that this annual death rate could double by 2020 (WHO 2002).

With an increasing global population comes an increase in demand for energy, food and water, caused also by an increasing standard of living in many developed countries and industrial expansion in developing countries (such as China and India). Without significant reductions in global greenhouse gas emissions, the effects of global warming will continue to impact on our everyday lives and potentially, and most importantly, our very existence. In recent years, the awareness of this challenge has grown. However, our current way of living, particularly in the Western world, must change if we are to avert the potential threats that are posed by climate change. Moderate reductions in greenhouse gas emissions will not be enough to combat and reverse the damage being done and that which has already occurred. An increasing world population (by approximately 80 million people per year (Central Intelligence Agency 2009)) and the further expansion of previously under-developed countries and the associated increasing demand for natural resources and energy that this brings all threaten to undermine any efforts being made to reduce global greenhouse gas emissions.

Collectively, we must strive for significant reductions in greenhouse gas emissions if we are to avert many of the predicted consequences of climate change. Hansen *et al.* (2008) suggest that the atmospheric carbon dioxide concentration levels must be at least as low as 350 ppm. Many leading scientists consider that long-term concentrations above this level would be potentially catastrophic (Hansen *et al.* 2008; Rockström *et al.* 2009). With current carbon dioxide concentrations at 387 ppm (Tans 2010), it is predicted that a reduction to 350 ppm would only be achievable through a global carbon dioxide emission target of at least 45 per cent below 1990 levels by 2020. This would result in only a 7 per cent probability of a 2°C temperature increase above pre-industrial levels rather than the 50 per cent chance based on an emissions stabilization level of 450 ppm (Meinshausen 2006).

1.2 Pollution

There are many human activities that produce pollutants which are released into the environment in large quantities, polluting the soil, air and water. These pollutants are released from modern energy-generation processes (coal-, gas- and petroleum-based); from refining and processing of raw materials to form construction materials (such as steel and cement); from industrial processes; and from the leaching of hazardous chemicals and contaminants from the waste disposed of in landfills, sewerage treatment plants and other waste disposal sites. These pollutants, particularly when in high concentrations, can have a significant impact on human health, the natural environment and local ecosystems.

The extraction of natural resources and many industrial processes are responsible for the release of a considerable quantity of these pollutants into the environment and significant contamination of waterways and soil. The release of chemicals and other pollutants into waterways may lead to significant eutrophication effects, whereby the concentrations of certain substances become higher than the natural balanced state (Smith *et al.* 1999). These concentrations of minerals and pollutants in water supplies can have considerable health implications for those that rely on this water for survival. The biological imbalance can also be detrimental to organisms and marine life that live in these waterways, causing indirect impacts on those that might rely on these for food (Asonye *et al.* 2007). For example, the 90 million tonnes of waste rock and tailings released annually from the Ok Tedi Mine in Papua New Guinea (OTML 2003), as well as the 1984 collapse of the dam system built to filter out much of the heavy metal contaminants, has caused significant contamination of the Ok Tedi River, surrounding waterways and agricultural land. This has in turn killed or contaminated the fish in the river system and left large quantities of contaminated mud on the land used to grow food consumed by the local population.

For many countries around the world, pollution of water sources represents a significant environmental concern. For example, in Australia nitrogen (particularly from fertilizers) represents the single largest quantity of emissions to water sources with over 32,000 tonnes released into the environment on an annual basis. The next four substances representing the greatest quantity of emissions released to water for Australia in the 2006–07 year are shown in Figure 1.5. This is also indicative of the main emissions of nutrients and other hazardous substances to water sources for many other parts of the world (see for example, European Environment Agency 2010).

The particulates released from industrial processes as well as other sources, such as the combustion of fossil fuels, are also a major source of air pollution and result in considerable human health impacts. In many industrialized cities around the world, smog produced from the concentration of significant quantities of car exhaust and industrial process emissions is visible evidence of the impact that human activities are having on the environment. This smog can cause respiratory problems in humans and has been linked to the increased occurrence of respiratory infections and asthma for people living in cities (WHO 2002).

The major source of pollutant emissions from human activity is the generation of electricity from fossil fuels. Of the harmful pollutants being released into the atmosphere and environment from electricity generation,

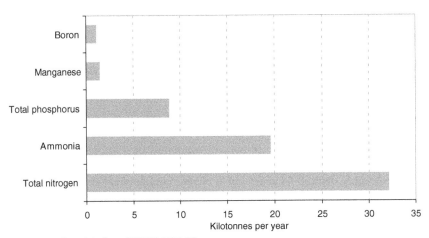

Source: Based on data from DEWHA 2009: 22.

Figure 1.5 Top five emissions to water for Australia, 2006–07

the release of sulphur dioxide to the air makes up by far the largest proportion, at 49 per cent (DEWHA 2010). Sulphur dioxide in particular can have significant impacts on land and water ecosystems by increasing the acidity of these environments as a result of acid rain. However, in industrialized countries, carbon monoxide emissions from fuel-reduction burning, wildfires and motor vehicles can also represent a significant proportion of total pollutant emissions (DEWHA 2010).

1.3 Resource depletion

The Earth contains an abundance of natural resources, including minerals, water, land and air. The value that we as humans place on these resources depends on their availability, accessibility, usefulness and our necessity for them. Fresh water, for example, is essential for maintaining human life and it is crucial to our very existence that we do not interfere with the Earth's ability to provide us with a clean, reliable supply.

Generally, these natural resources are used to sustain human life (water and food), for energy generation (coal, natural gas and petroleum) or to produce more complex products or materials (iron ore, petroleum, bauxite, limestone, timber etc.). Many of these resources are unable to be naturally replenished by the Earth at anything like the rate at which humans are currently consuming them and are thus considered to be 'non-renewable' or 'finite'. Once these non-renewable resources have been exploited, or at the point where they become considered too costly or difficult to extract from the Earth, we will have no choice but to look for or develop alternative solutions to meet our needs. The wasteful and excessive consumption of natural resources is a habit that was developed at a time when resources were considered to be in abundance. Reserves at the time were not thought of as being in danger of exhaustion. More recently, increases in population and per capita consumption have brought us much closer to the point where the depletion of many of the Earth's resources is imminent.

Human activities, such as the construction and operation of buildings and other human-made infrastructure and the manufacture of consumer goods, consume considerable quantities of natural resources. For example, globally over 2.2 billion tonnes of iron ore is extracted from the ground on an annual basis (US Geological Survey 2009: 81), primarily to produce steel that is used in constructing the world's buildings, bridges and other infrastructure. Total current accessible iron ore reserves are estimated at 150 billion tonnes (US Geological Survey 2009: 81) and it is estimated by Brown (2006: 109) that this equates to only another 64 years worth of reserves based on an annual growth rate in extraction of 2 per cent. This figure could be seen as extremely conservative considering that continuing demand from China and India in particular are likely to drive growth in iron ore extraction rates well beyond this.

As the world's population increases and many previously under-developed countries (such as China and India) invest heavily in infrastructure, demand for natural resources continues to grow. Because natural resources are used to make many of the things that we consume, population and consumption are intrinsically linked and inevitably lead to an increase in total resource consumption. Resources already facing imminent depletion are at even greater risk of being exploited much sooner than has been previously predicted. This makes the need to shift to a more sustainable level of consumption and way of living even more urgent if we are to have any hope of maintaining the planet in a state that is liveable for future generations.

The extent of the impacts associated with the depletion of resources is dependent on whether or not we are able to find alternative means of meeting human needs before these resources are depleted. If our rate of consumption continues on its current path, and alternative resources or technologies are not found in time, human civilizations may face a resource, economic and social crisis. Many of the easily and economically accessible non-renewable resources on which we heavily rely are predicted to be exhausted within the next 100 years (CIRCA 2008). For example, it is predicted at current consumption rates that there is less than 50 years of proven oil reserves remaining (CIRCA 2008; BP p.l.c. 2010). However, with consumption increasing, these reserves too may in fact be depleted much sooner than these figures might suggest. Other than the potential consequences associated with exhausting the Earth's resource reserves, there are also considerable environmental impacts that result from the mining, harvesting and resource-extraction processes. Ecosystems are destroyed, fossil fuels are consumed by machinery and for transportation and associated greenhouse gases are released into the atmosphere, waste is generated and water sources are contaminated.

Land is also one of the most valuable resources available on Earth as it is essential for providing food for human consumption, for harnessing renewable energy (solar, wind, biomass etc.) and growing timber. The destruction or depletion of this land resource is occurring through the contamination of soils, salination, depletion of nutrients through over-farming, acidification and its general over-use in response to providing housing and food for the world's expanding population.

WORLD POPULATION

The global population has grown exponentially over the past several centuries, doubling in the past 40 years alone to 6.8 billion people (Central Intelligence Agency 2009) (Figure 1.6). Currently, the global population growth rate is 1.18 per cent per annum, equivalent to 220,980 people per day. Even if significant reductions in resource consumption and greenhouse gas emissions are achieved, the rate at which global population is increasing (particularly in developing countries such as India and China) will mean that any efficiency improvements, cleaner production processes and conservation strategies may be offset by the increasing demand for global resources, energy and water from this rapidly growing population. Herein lies a considerable opportunity to re-think how we might meet the needs for shelter, food, energy and water, of both current and future generations.

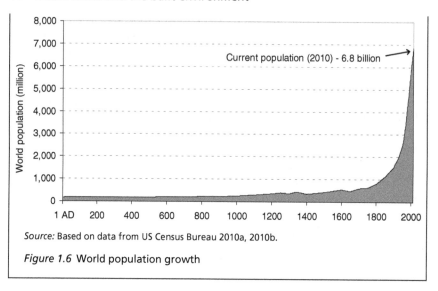

Source: Based on data from US Census Bureau 2010a, 2010b.

Figure 1.6 World population growth

1.4 Production and disposal of waste

Human activities generate enormous quantities of waste that is often incinerated or discarded in landfill sites, with a small proportion of this recovered for other uses (for example, as a fuel for industrial processes or as a raw material for new materials or products). As countries become more developed, standards of living and demand for goods and services rise. This, together with a rising population, generates an increasing quantity of solid and hazardous waste that must be disposed of or treated.

Waste production occurs across every sector, from mining, agriculture, manufacturing, energy production, construction and within households. The proportion of waste disposed of in landfill varies considerably between countries. Waste disposal and storage is a much greater issue in countries where land is at a premium. Disposing of waste in countries where available land is, or will one day become, much more scarce can have implications for competing land uses, such as food production, energy generation or housing. In areas such as these, waste is often incinerated, releasing greenhouse gases such as methane and carbon dioxide as well as other potentially dangerous gases and toxic substances into the atmosphere. This is the case in many European countries and countries where land is in high demand for other uses (such as Japan).

In countries such as Australia and Ireland, where undeveloped land is in much greater abundance and therefore the value placed on it is not often as great as in some other countries, incineration of waste rarely occurs. Figure 1.7 shows the proportion of produced municipal waste that is sent to landfill, incinerated or recycled in selected OECD countries.

Soil, building rubble (such as bricks and concrete from demolished buildings), food scraps and garden waste make up the largest proportion of total waste

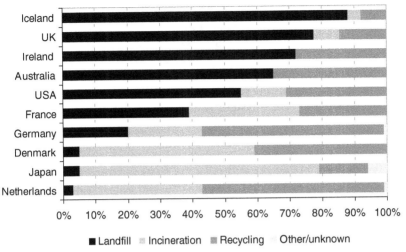

■ Landfill ▨ Incineration ▨ Recycling ▨ Other/unknown

Source: Based on data from OECD 2005.

Figure 1.7 The treatment of municipal waste in selected OECD countries, 2003

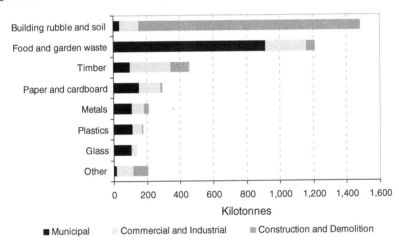

■ Municipal ▨ Commercial and Industrial ▨ Construction and Demolition

Source: Based on data from DEH 2006.

Figure 1.8 The composition of waste disposed to landfill in Victoria, Australia, 2002–03

that ends up in landfill in many developed countries, as evidenced for the State of Victoria, Australia in Figure 1.8.

The waste produced from the construction and demolition (C&D) sector often represents the largest proportion of generated waste in many countries. For example, in Australia C&D waste represents 42 per cent of generated waste, of which 43 per cent is disposed of in landfill (DEH 2006: 8). Municipal (27 per cent) and commercial and industrial waste (29 per cent) account for the remainder (DEH 2006: 8; Productivity Commission 2006).

In light of a growing awareness of the potential environmental impacts of waste production and disposal, significant attempts are being made across the world to minimize these impacts by diverting waste materials from landfill to be reused or reprocessed for other uses or in new materials and products. Waste recovery minimizes the requirement for landfill and incineration, and their associated environmental impacts, and reduces demand for virgin raw materials. In many countries, this rate of waste recovery has been increasing over the past decade as new sorting processes and reprocessing technologies are developed. The rate of waste recovery varies considerably by country, ranging from less than 10 per cent (e.g. in Iceland) to close to 60 per cent (e.g. in countries like Germany and The Netherlands) of all waste produced (Figure 1.7). As an example of the rate at which waste recovery has been increasing over recent years, in Victoria, Australia, the waste recovery rate has increased from 40 per cent to 62 per cent between 1997 and 2007 (Figure 1.9). The most significant increases in recovery rates have been seen where the cost of waste disposal by traditional means has increased (directly through additional taxes or indirectly through the higher value put on land, for example). This has meant that the value of certain waste materials has increased through a closing of the gap between the costs of traditional disposal methods and typically more cost-intensive recycling practices.

Where the waste associated with the construction and demolition (C&D) of buildings and built environment infrastructure is recovered for recycling, concrete can typically account for at least half of the recovered materials by weight (Figure 1.10) (Sandler 2003; Sustainability Victoria 2008).

While the generation of waste is not a direct environmental impact in itself, the disposal and treatment of this waste can have significant environmental

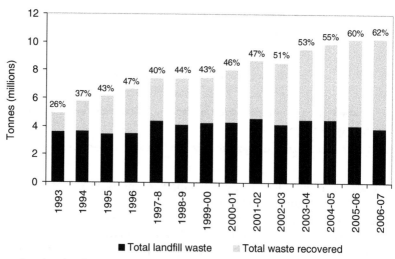

Source: Based on data from Sustainability Victoria 2008.

Figure 1.9 Waste generation and recovery rates for Victoria, Australia, 1993 to 2006–07

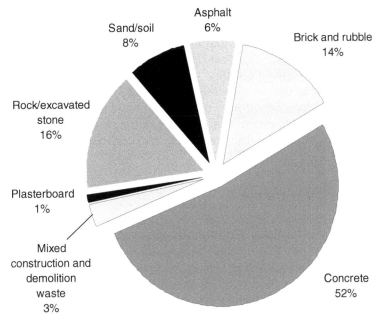

Source: Sustainability Victoria 2008.

Figure 1.10 Composition of construction and demolition waste (by weight) recovered for reprocessing in Victoria, Australia, 2006–07

consequences. A considerable proportion of the waste disposed of in landfill is inert and one of the most obvious environmental consequences of its disposal is the land that this takes up. Even more harmful to the environment is waste that contains toxic or hazardous substances or high concentrations of chemicals or heavy metals. These substances can leach into the surrounding environment, causing long-term and potentially irreversible changes to local ecosystems, contaminating water supplies and potentially poisoning humans and wildlife. Many of the organic materials disposed of in landfill sites also release large quantities of greenhouse gases, such as methane and carbon dioxide, as they decompose. These gases are either stored in the soil or released to the atmosphere where they may contribute to global warming. The incineration of waste can also have significant environmental and human health consequences through the emission of hazardous gases and particulate matter that are released during the process, although filters can be used to help minimize or prevent the release of these contaminants into the environment.

Despite a gradual increase in the rate of waste material recovery worldwide, the quantity of waste that is generated continues to increase, as shown in Figure 1.9 for Victoria, Australia. This increase in waste production is due to both population and economic growth and it is predicted that this growth will continue into the future (OECD 2008). This increase in waste generation is a direct indication of an increasing rate of resource consumption. Like many of our current attempts to reduce the environmental impacts from human activities, our current efforts at waste reduction are at best only slowing down the problem, rather than resulting in any net reduction in the quantity of waste being generated. Significant improvements and changes in current design practices, material selection, construction techniques and materials-recovery processes will be necessary if we are to considerably reduce the impacts on the environment from the treatment and disposal of waste. Even then, the billions of tonnes of waste already existing in the environment will continue to impact on our lives and the Earth's ecosystems for many decades to come.

1.5 The built environment

The built environment includes a diverse range of human-made infrastructure systems, including buildings, transport networks (roads, bridges, railways etc.) and utilities (water, power, telecommunications etc.). These systems are considered necessary for human survival, particularly in urban areas inhabited by large populations. Without these systems, large cities would cease to function and human health and survival would be significantly jeopardized.

Buildings are by far the most complex element within the built environment, often involving a combination of dozens of different materials, components and systems. They must provide a comfortable environment for human occupants and serve a broad range of functions (housing, offices, factories, hospitals, education facilities and so on). Throughout history, buildings have adapted to climatic conditions, for functional reasons or for aesthetic and cultural reasons, due to the availability of materials and due to technological advances, but their primary purpose is to provide shelter from the weather for their inhabitants.

The construction of any human-made structure requires some form of natural resources from the Earth. Buildings also require resources to control human comfort, such as energy for heating, cooling and lighting, and to provide basic essential services such as pipelines for the provision of water and disposal of waste. With these requirements for natural resources comes a range of environmental impacts, which occur across the life of a building.

It is estimated that, worldwide, the operation of buildings alone accounts for up to 40 per cent of total global energy demand, resource consumption and greenhouse gas emissions (OECD 2003). This excludes the resources, energy and emissions associated with the manufacture of materials and construction of buildings, which can be as significant as the operational demands. Building materials are sourced from a wide range of industries, where manufacturing processes often result in a large consumption of resources and energy and release of emissions. This consumption and these emissions, and their related environmental impacts, can be directly attributed to the construction of buildings, because without the demand for these resources by the construction industry, the associated impacts would not otherwise occur.

Buildings are also responsible for a considerable proportion of global water consumption, waste production and release of pollutants to the environment. For example, it is estimated that around 12 per cent of global fresh water consumption is used within buildings (US Geological Survey 2005). In addition to this, significant quantities of water are also required during the manufacture of building materials and during the construction process (for the production of concrete, for example).

1.5.1 The built environment life cycle and its related environmental impacts

Systems or elements within the built environment, such as buildings, typically last many decades and sometimes many hundreds of years. During this time, these systems go through a range of life cycle stages from the extraction or acquisition of the raw materials required for their initial construction through to their eventual demolition and disposal, as depicted in Figure 1.11 for buildings.

Environmental impacts occur across each of the different stages of the built environment life cycle and relate to the extraction of raw materials from the earth, processing and manufacturing of building materials, the construction process, operation or use, maintenance and repair, refurbishment and eventual disposal or reuse of materials. Natural resources, energy and water are consumed, greenhouse gas emissions and other pollutants are released into the environment and waste is produced across each of these stages. The

Figure 1.11 Stages involved in the building life cycle

environmental impacts resulting from these resource requirements and the production of emissions and waste vary both temporally and spatially and are dependent on a range of factors, such as production methods, source of raw materials and fuel types used, to name only a few.

1.5.1.1 Raw material extraction

All elements within the built environment, such as buildings and infrastructure systems, require raw materials from the earth to make the materials used in their construction. These raw materials (such as iron ore, limestone, bauxite, copper, timber and petroleum) are extracted from the earth and processed using a variety of technologies into more useful forms to be used as building or construction materials (such as steel, cement, aluminium and plastics). The continued extraction of non-renewable resources such as iron ore and petroleum will eventually lead to their depletion, unless alternative solutions for meeting our resource demands are found (Meadows *et al.* 1972; Daly 1990).

The extraction of raw materials from the earth not only results in the eventual depletion of those natural resources but also involves the consumption of considerable quantities of energy and water as well as the release of emissions and pollutants. Mining operations rely on fossil-fuel-based energy to power equipment and machinery that is used during the mining process and also to operate the transport equipment and infrastructure necessary to transport these raw materials to their place of processing or use. Environmental impacts result from the production and use of this fuel in the form of greenhouse gas emissions and release of other pollutants, such as carbon monoxide, into the atmosphere.

The environmental impacts of mining continue long after the mining process has taken place. Tailings are a waste by-product of the mining process and their disposal can have significant environmental consequences, potentially contaminating the surrounding environment, including local water supplies, with unnaturally high concentrations of chemicals and minerals. Many mining operations use chemicals to assist in the separation of valuable ores from the waste materials and these chemicals remain as residues in the tailings materials for many years. Other impacts of the mining process include erosion and a loss of biodiversity, particularly where large areas of land are cleared for open-cut mines or in harvesting timber from forests.

1.5.1.2 Manufacturing

RAW MATERIAL PROCESSING AND BASIC MATERIAL MANUFACTURE

This stage of the built environment life cycle involves taking the raw materials and converting them into basic building materials, such as steel and cement. The manufacture of construction materials involves many complex processes. For example, iron ore extracted from the ground undergoes a number of chemical processes and is combined with a number of other elements (such as carbon) to form steel.

The raw material processing stage is one of the most energy-intensive processes involved in the manufacture of many materials. For example, the energy required to produce cement represents at least 65 per cent of the energy required to manufacture concrete (Zapata and Gambatese 2005). Also, the electric arc furnaces used in steel making require significant quantities of electricity to operate (Jones *et al.* 1998). Most often, this energy is fossil-fuel-based and its production releases significant quantities of greenhouse gases into the atmosphere. These processes can also generate large quantities of waste materials that, if not managed appropriately, can also contaminate the surrounding environment.

MANUFACTURE OF CONSTRUCTION COMPONENTS

Materials such as steel and concrete in their basic form, and often combined with other materials, are then used to manufacture construction elements such as structural supports (for example, reinforced concrete columns and beams), cladding, finishes and services that are used within the built environment. Sometimes the processing and product manufacturing stages are combined and the manufacture of the finished product follows immediately after and is actually part of, the raw material processing stage (e.g. for glass and bricks where the basic raw material, after being produced, is immediately formed into the finished product).

The manufacturing stage may involve further material processing (for example, sawing of timber into the appropriate shapes and sizes required for construction), the addition of other constituents (aggregates and water added to cement to make concrete) or forming and fabrication (for example, turning basic steel in roll form into corrugated roof sheeting products). The manufacturing stage may also involve a level of prefabrication of built environment components. Products made from a number of different materials, such as windows, pre-cast concrete or composite wall panels, may require individual materials to be fixed to one another to form the finished product before the construction process commences.

As for the materials processing stage, the processes involved in converting basic materials into more complex and useable products and components can have a considerable impact on the environment. Industrial processes typically require significant quantities of energy and water. Whilst some industries have taken a proactive approach to utilizing energy from renewable sources, the majority of this energy is fossil-fuel-based, generating greenhouse gas emissions and releasing other pollutants into the environment. Industry is also beginning to become much more proactive in conserving water resources, a move that has been heavily influenced by the current drought conditions in many parts of the world, particularly in Australia where it is estimated that recent improvements in water efficiency, including increased reuse or recycling of process water and the collection and use of rainwater, has resulted in up to a 41 per cent reduction in industrial fresh water demand (DSE 2010). The water used during manufacturing processes typically draws upon resources used for human consumption. In countries like Australia, a large proportion of this water is recycled; however, the treatment of this water for recycling or disposal has an impact on the environment through the chemicals that are used in these treatment processes.

The majority of the considerable quantities of waste often generated by manufacturing processes have traditionally been disposed of in landfill. However, as resource values continue to increase, much of this waste is now being recovered for reuse in the manufacturing process.

1.5.1.3 Construction

Manufactured building elements are combined on site during the construction stage to form a part of the built environment, whether it is a building or other infrastructure system. Various technologies and processes are used during this construction stage which in itself results in a range of additional environmental impacts. As for every other stage of the built environment life cycle, energy and water are consumed and waste is produced.

The transportation of building materials and components to the construction site is generally performed by trucks which require fossil fuels to operate. On-site construction processes require energy, usually in the form of electricity, to power electrical tools and operate machinery and other equipment. The construction process can also consume considerable quantities of water, mainly for cleaning, concrete curing and dust suppression. Typical construction processes also generate a considerable quantity of waste due to human error, accidental damage, off-cuts and the common practice of ordering surplus materials to allow for wastage. A significant proportion of these waste materials is traditionally sent to landfill and can contribute to soil and water contamination on these sites. However, a considerable quantity of the waste materials that would have otherwise been sent to landfill is now being recovered and recycled. This is highly evident in countries such as Australia, where around 57 per cent of construction waste is recycled (DEH 2006).

1.5.1.4 Operation and maintenance

OPERATION AND USE

Once elements of the built environment are constructed, they enter their operation or use stage. The different components of the built environment have varying requirements and impacts during this stage. One of the basic requirements for the operation of many components of the built environment is energy. Energy is required to power lighting systems and control signals for road and rail infrastructure, to power pumps and control systems for distributed water, electricity and telecommunications networks and to provide heating, cooling, lighting, hot water and power to buildings. Generally, this energy is provided from fossil-fuel-based sources. In recent years there has been a gradual increase in the use of solar electric systems for the provision of lighting in public places and to provide the energy required to operate buildings. However, worldwide, 86 per cent of energy consumption is still provided by fossil fuels (EIA 2008a) and this consumption results in approximately 30,000 Mt of greenhouse gas emissions per annum (EIA 2008b). There have been concerted efforts to improve the energy efficiency of building operation for many years now. These efforts focus on improving thermal performance, a major component of passive solar design where buildings are designed and oriented to maximize the benefits of local climatic conditions; public awareness campaigns to direct behavioural change; and improving the energy efficiency of appliances and systems within buildings through design and technological development.

Water use in buildings can also be significant, particularly for the operation of cooling systems and for cleaning, washing and drinking. Considerable effort, particularly in countries experiencing severe drought conditions, has been placed on improving the water efficiency of building operation. While energy-efficient design and appliances have been a focus for building designers for a number of years now, the efficient use of water within buildings is a much more recent concept. Appliances with greater water efficiencies, low-flow taps and waterless urinals all help to reduce the water consumption and associated environmental impacts of buildings. Systems that treat and recycle the water that is used are also being integrated into many building projects. These systems maximize the use of the valuable water resource, whilst minimizing reliance on fresh water. Many recent building projects around the world – such as the City of Melbourne's new office building, Council House 2 (CH_2) – have shown that it is possible to reduce mains water consumption by as much as 72 per cent (City of Melbourne 2009).

Another important environmental issue relating to the operation of the built environment, and particularly buildings, is the quantity of waste that is generated during this stage. This includes human waste, food waste, paper, packaging and other wastes associated with the particular practices occurring within the building that may be part of specific manufacturing or industrial processes. For typical office buildings, for example, over 50 per cent and up to 72 per cent of this waste is paper which is either sent to landfill or recycled into new paper-based products (ABC 2010; WasteCap 2010). When sent to landfill, the anaerobic decomposition of this waste paper can result in soil and water contamination through the leaching of chemicals and other pollutants used in its production.

MAINTENANCE AND REPAIR

This stage of the built environment life cycle involves the maintenance of materials, systems and components of buildings or infrastructure in order for them to continue to perform or operate in the safe, efficient and reliable manner initially intended. Maintenance also involves periodic cleaning and replacement of consumable components such as air conditioning filters and light fittings. Repairs are carried out during the life of these systems as their necessity arises due to damage or deterioration. Re-painting is generally performed on a regular basis, particularly to protect weather-exposed elements in order to prolong their life and physical appearance. Regular inspection of buildings and infrastructure systems is a necessity to ensure that their components do not become beyond repair. This can lead to a reduced life expectancy of these components or even whole buildings or systems and even catastrophic failure: for example, where structural integrity or functionality is compromised. Insufficient maintenance regimes may not only have safety and operational implications for the occupants of the built environment; they may also lead to increased environmental impacts through the replacement of components earlier than anticipated, thus consuming more resources than would otherwise have been necessary.

Damaged or replaced materials and components must be disposed of, often in landfill. Items such as light fittings can contain hazardous chemicals, including mercury, that when disposed of in landfill can contaminate the surrounding environment and water sources. The less durable these components are, the more frequent is their replacement and the greater the amount of waste generated. New components must also be used to replace those that have been discarded, and the manufacture of these results in even further demand for resources and environmental impacts. In addition, hazardous chemicals are often used for cleaning buildings. These chemicals can make their way into the environment, having serious implications for human health and local ecosystems (Goodman 1974).

REFURBISHMENT

Building or infrastructure refurbishment often occurs once a building or infrastructure system is considered to have come to the end of its useful life or its individual elements are considered obsolete or beyond repair. Refurbishment allows a building or infrastructure system to be kept or reused rather than demolished. Refurbishment of buildings is common where the structural integrity of the main structure remains but technological, functional, safety or market developments warrant changes to the building fabric, services or finishes. The refurbishment of buildings ranges from very minor – such as replacement of carpets, re-painting and re-configuration – to major, such as extensions, conversion to an alternate use or completely stripping the building back to its main structure and re-cladding and re-fitting it.

As with initial construction, environmental impacts during the refurbishment stage result from the manufacture and installation of materials and components that are used to replace deteriorated or unwanted elements. Impacts include a depletion of those natural resources used; emissions from processing, transportation and energy consumption; water consumption; and waste generation and disposal during the manufacturing processes and on site.

In addition, the materials that are being removed from the building need to be disposed of, which requires fuel for transportation and may result in soil and water contamination when these materials are put into landfill or an additional demand for energy and water if they are re-processed into new materials. The refurbishment of buildings can produce significant quantities of waste; however by keeping those materials and components that are able to be retained, the life of these materials and components is able to be extended, less material is sent to landfill and the need for additional virgin resources can be minimized, thereby avoiding many of the impacts associated with the extraction, processing and manufacture of replacement materials.

1.5.1.5 End-of-life (demolition, disposal, reuse and recycling)

The final stage of the built environment life cycle is the demolition or dismantling of the building or infrastructure system and the subsequent disposal or reuse of individual components and materials. Buildings and infrastructure systems have a finite life and eventually come to a point where they are no longer considered suitable or safe for human use or occupancy. The deterioration of materials beyond repair, increase in the number of users, technological advancements and general market expectations all contribute to determining the effective useful life of a building or infrastructure system (Langston *et al.* 2008). Once at the end of their useful life, these systems are often demolished or dismantled to make way for new buildings and infrastructure that is safer or more suited to current demands, needs and expectations.

Generally, the more durable and long-lasting the materials and components that make up a building or infrastructure system are, the lower the environmental impacts through reductions in replacement, maintenance and refurbishment. However, one exception to this may be where an over-specification of materials occurs due to their physical life far exceeding their intended life. For example, the constant churn of furniture and fittings, such as carpet in office buildings, may create greater environmental impacts when highly resource-intensive materials are specified for their durability instead of more appropriate options that are more suited to the (shorter) intended life of those elements.

The environmental impacts associated with the end-of-life stage of the built environment life cycle include those associated with the demolition and material removal processes, including emissions from the fuel used for demolition machinery and transportation of waste to landfill or materials to recovery sites. When these materials are disposed of in landfill they may leach chemicals into the ground, potentially resulting in eutrophication of the surrounding environment. As organic materials, such as timber, decompose they may also release carbon dioxide and methane into the soil and atmosphere. While the recovery and reuse of existing materials may avoid or reduce some of the impacts of using new materials, the reuse of these materials is not without certain impacts of its own. Where these materials need to be re-processed, fuel is required for transportation and energy and other resources are required for re-processing, which themselves result in additional environmental impacts.

1.6 Summary

Human activities result in a broad range of impacts on the natural environment, from the depletion of precious natural resources to global warming and the pollution of the air, land and water on which we rely for our very survival. Many of these impacts are the result of the construction and use of the built environment, including the buildings and various infrastructure systems considered essential to support the large number of people inhabiting the towns and cities of the world. These impacts are spread not only across the different areas of the natural environment, but also across many decades. Decisions made today are often locked in for generations and can have significant long-term and lasting implications for human and environmental health for many years into the future. The built environment involves numerous stages from the extraction of raw materials to construction, operation, replacement and eventual disposal or recovery, all resulting in a variety of environmental impacts that continue long after buildings and other infrastructure systems are no longer in use. From this stems a number of issues for built environment designers, developers, owners, occupiers and managers.

Current consumption levels, particularly in the developed world, are not sustainable. Many of the Earth's natural resources are being depleted at a much greater rate than they are able to be renewed. In addition, our current pollution of the natural environment may be severely detrimental to not only the fragile ecosystems of the Earth, but also our very survival. If we are to continue to enjoy or improve our current standards of living, we must take a much more sustainable and less polluting approach to our activities. We must look to more efficient and less polluting forms of energy production, find alternatives to the non-renewable natural resources we currently consume, develop cleaner processes for production, and reduce and find better ways of treating the waste that we produce. One of the greatest challenges that human society is currently facing is how to support the growing global population with a rapidly depleting natural resource base. An increasing population also threatens to negate any environmental impact mitigation strategies if associated resource consumption and pollution continue to increase at a greater rate than the improvements being made by such strategies.

The built environment is one area in which considerable potential exists to make large improvements in environmental performance and reduce many of the impacts caused by human activity. Because of the huge quantities of energy, water and resources consumed and pollutants and emissions produced from the construction and operation of the built environment, there is substantial scope to reduce the impacts that occur as a result of these processes. Creating a more sustainable built environment will go a long way to ensuring the future security and prosperity of human civilization.

2 Towards a sustainable built environment

With the increasing adverse impacts that the built environment is having on the natural environment and on human health, the need to address these impacts and look for alternatives, remedies and solutions to the current ways in which our buildings and infrastructure are designed and managed is becoming more and more crucial. Around the world, there is a growing emphasis on reducing the impacts that we as humans are having on the natural environment. Whilst this awareness of environmental issues has been with us for many decades, the widespread acceptance and desire to act on broad environmental concerns really only began in the 1990s. Much of the concern stemmed from the dire predictions of potential catastrophic consequences from human-induced climate change. However, climate change impacts are not the only focus of those looking to improve the environmental performance of the built environment and human activity. Whilst causes of climate change, such as the release of greenhouse gases from fossil fuel combustion and agriculture, have been the focus of many governments and groups around the world, water and soil contamination, water consumption, land degradation, waste treatment and disposal and the depletion of natural resources are other important areas where an increasing emphasis is being, and should be, placed.

As stated in Chapter 1, our current resource-intensive and highly environmentally adverse approach to meeting human needs is not sustainable. This is becoming even more evident considering even conservative predictions of human population growth over the coming years and the resultant increasing demand that this will place on these resources. Reducing the impacts associated with the design, construction, operation and management of the built environment will take a concerted effort on the part of the many stakeholders involved across all of the stages of the built environment life cycle. In order to reduce these impacts and encourage a more sustainable approach within the built environment we must develop a detailed and rigorous understanding of the complex and varied interactions that exist between the natural and built environments. Only then can we hope to achieve a built environment that is sustainable and does not in any way compromise the very existence of future generations. The beginning of this understanding involves the identification of the environmental impacts that are caused by our current construction processes, energy sources and waste management practices. This process is essential in order to target improvement strategies towards the areas where the most adverse impacts are occurring.

Assessment of the potential environmental impacts from the built environment has been occurring for many decades. Typically, the desire to gain an understanding of these impacts has been a direct result of global economic, environmental or social concerns at various points in history. This chapter provides an overview of what must be done to minimize or avoid many of the adverse impacts that the built environment currently has on the natural environment. The concept of *environmental design* will be explained as an integral part of this process, with particular emphasis on its application in the built environment. Following this, the origins of *environmental assessment* and the changing focus and purpose of its use in the built environment are described.

2.1 Minimizing the environmental impact of the built environment

Chapter 1 demonstrated some of the many impacts that the built environment may have on the natural environment as a result of the consumption of resources, release of pollutants and the production and disposal of waste. A vast range of opportunities exists for reducing, or even avoiding, many of these negative impacts. These opportunities are often greatest and best able to be captured during the design process, either prior to initial construction for new buildings and infrastructure systems or in the redesign or redevelopment of existing buildings, components and systems. *Environmental design* provides a basis for identifying and capturing many of these opportunities and a framework for encouraging the development of a more environmentally sustainable built environment.

2.2 Designing for the environment: strategies for a sustainable built environment

There are numerous strategies that can be employed or considered during the design process in order to reduce the adverse environmental impacts of the built environment. Many of the principles for improving the environmental impact of the built environment are addressed within the *environmental design* process.

Environmental design, which is also known as *sustainable design, ecodesign* or *Design for Environment (DfE)*, involves the consideration of the life cycle environmental impacts of a product or process during its design. Broadly, this involves minimizing or avoiding *demands* from nature that have an adverse impact (either short or long term) on the environment and minimizing or avoiding *outputs* to nature that have an adverse impact on the environment. For maximum benefits to be achieved from environmental design, it must involve all components and participants in the extended supply chain, including not only the specification of more resource-efficient materials and systems, but also the development of more environmentally benign materials and improvements to manufacturing processes. Limitations to achieving optimal environmental outcomes are often imposed by the availability of alternative solutions. Therefore, the limited ability for design professionals to minimize environmental impacts should be acknowledged. Deeper resource savings and environmental goals can only be achieved through a more integrated approach that involves governments, consumers, designers, manufacturers and suppliers.

A number of the environmental design principles that are relevant to the built environment are described in the following sections. A summary of what can be done during the design process to improve the environmental performance of the built environment is shown in Table 2.1.

Table 2.1 Strategies for improving the environmental performance of the built environment

Strategy	Environmental benefits
Use resources more efficiently	• Preservation of non-renewable resources • Sustainable consumption of renewable resources • Reduced waste production
Minimize non-renewable resource use	• Preservation of non-renewable resources • Minimized emissions from energy production
Minimize pollutant releases	• Maximized water, air and soil quality • Preservation of ecosystems
Design for disassembly	• Preservation of natural resources • Maximized resource value • Reduced waste production
Minimize waste production	• Minimized low-value land activities (i.e. landfill) • Minimized soil and water contamination • Maximized resource value
Design for recyclability	• Preservation of natural resources • Maximized resource value • Reduced waste production
Design for durability	• Reduced demand for raw materials, energy, water • Preservation of non-renewable resources • Reduced waste production
Design for adaptive reuse	• Preservation of natural resources • Reduced demand for raw materials, energy, water • Maximized resource value • Reduced waste production

2.2.1 Resource efficiency

The efficient use of resources involves maximizing the value that resources provide whilst minimizing the environmental impacts associated with their use. Using existing resources such as fossil fuels, raw materials, water and land more efficiently can help to slow resource consumption, which is particularly crucial for those resources considered to be non-renewable. Improving the efficiency of resource consumption in the built environment has traditionally been focused on the operational efficiency of buildings. This includes improving the thermal performance of building envelopes to reduce demands for artificial heating and cooling by designing according to passive design principles, considering the correct orientation, maximizing direct solar gain during cooler periods and minimizing it during warmer periods; replacing lighting systems with more efficient fittings; and installing more energy- and water-efficient appliances. Significant resource savings can also be achieved by improving the efficiency of the manufacturing industries that supply materials and products to the construction industry (Sorrell *et al.* 2004). Upgrades to manufacturing equipment can improve resource efficiencies by utilizing systems that operate more efficiently, consuming fewer raw materials and less energy and water. Waste heat from manufacturing processes (such as from the kilns used to fire clay bricks or the furnaces used to produce steel) can also be captured to be reused for other purposes within the manufacturing plant.

2.2.2 Minimizing non-renewable resource consumption

The use of finite or non-renewable resources leads to their eventual depletion and ultimately the need for alternative resources and approaches to meet the many human needs that such resources currently provide. Development of new technologies, materials and processes takes time and waiting until the point at which the Earth's finite resources have been depleted or their extraction and use is no longer viable may lead to further social and economic problems, not to mention the damage caused to the environment in the process. For this reason, minimizing the consumption of non-renewable resources goes hand-in-hand with the need to find alternative resources or approaches to meeting the needs that they provide in order to avoid further environmental damage from current resource extraction, energy generation and material production processes. Fossil-fuel-based energy systems must be replaced with more renewable approaches to energy supply (such as solar and wind), whilst finite minerals and raw materials must be replaced with renewable alternatives. In addition, the technologies and current practices that rely on these resources may also need to be adapted or even replaced (the internal combustion engine is a good example of this from the automobile sector). We must also ensure that naturally renewable resources, such as timber and water, are used sustainably to ensure that they too are not depleted by overuse, especially in light of the growing demands for these resources from an expanding global population.

Using materials that are local and require minimal processing can reduce the demand for transportation and its associated impacts (from fuel consumption, for example) and help to avoid the often resource-intensive manufacturing processes associated with some of the most common construction materials. Significant reductions in non-renewable resource use can also be gained by rationalizing the design of buildings. Reducing the size of buildings as much as possible – especially where the space provided is considered to be excessive or more than is realistically required, as is currently the case in parts of many developed countries (such as Australia and the United States) – can result in considerable resource savings, not only in the embodied energy in the materials used to construct these buildings, but also in the energy needed to operate them and manufacture the furniture, appliances and other fittings needed to fill the additional space (Fuller *et al.* 2009).

The use of recycled or recyclable materials can also help to alleviate some of our reliance on non-renewable resources. However, as demand for raw materials continues to grow, so too does the demand for recycled materials, which can outstrip actual supply, as is beginning to be seen for timber in some parts of the world. Care should also be taken to ensure that the use of recycled materials does not sacrifice the overall performance of a building or infrastructure system by reducing its durability, safety, financial viability, service-life or operational efficiency. Design for recyclability and disassembly are important considerations in maximizing the opportunities for using recycled materials.

2.2.3 Minimizing pollution

The environment has a limited capacity to deal with and recover from a certain level of contamination (usually over the longer term); however, the current rate at which our atmosphere, soils and waterways are being polluted is well beyond the Earth's ability to reverse the damage that is being, and has already been, done. This means that every effort should be made to minimize the release of pollutants (including chemicals, heavy metals and greenhouse gases) into the environment, especially those considered to be toxic, to avoid potential further contamination. Every stage of the built environment life cycle is responsible for pollution of the natural environment, from the tailings that are produced from many raw material mining processes (such as metal ores), through to the chemicals used and released from some manufacturing processes and the leaching of substances from materials when they have been disposed of in landfill sites. Minimizing the release of these and other pollutants can help to ensure that the air, water and soil quality is maintained and also minimizes potential impacts of their release into the environment on human health.

Strategies for minimizing the production and release of these pollutants include eliminating the use of construction materials that result in pollutant releases at any stage of their life cycle. Where potential sources of pollutant releases exist, the use and disposal of these pollutants must be carefully monitored and controlled. Environmental and waste management plans are an essential element of this process.

The principles of *cleaner production* are also an important component of minimizing pollution. This is where industrial production processes produce less pollutant waste by monitoring outputs, reusing or substituting raw materials and implementing improvements or changes in organizational practices and technologies.

2.2.4 Designing for disassembly

Designing the built environment so that its materials, systems and components can be easily pulled apart or disassembled at the end of their physical or useful life has the ability to significantly enhance its long-term environmental performance by encouraging reuse and recycling, minimizing waste production and providing greater opportunities for adaptive reuse.

There are a number of key strategies for encouraging the disassembly and subsequent reuse of materials and components within the built environment. These include:

- using fastening and joining techniques that facilitate ease of disassembly (e.g. fasteners rather than adhesives; screws and bolts instead of nails and welding)
- minimizing the weight of individual parts
- avoiding the use of composite materials where separation of individual materials is difficult.

Whilst many of the materials that are currently able to be reused are not recovered, a high proportion of some highly resource-intensive materials such as bricks, steel and concrete are recovered for reuse or recycling. There is also considerable scope to expand this to other materials, especially if they are designed to be disassembled with limited cross-contamination from other materials. Designs that facilitate the separation of waste materials from reusable or recyclable materials can also reduce disposal costs.

2.2.5 Minimizing solid waste production

Some of the greatest environmental impacts of the built environment are those associated with the waste generated from the production and eventual disposal of construction materials. The production of waste from every stage of the built environment life cycle must be reduced and any waste that is produced must be carefully and appropriately managed so as not to have any long-term impacts on the environment. Current construction practices inevitably lead to the production of waste. Where this is unavoidable, after all attempts at its minimization, every effort should be made to recover these materials and to reintroduce them into the supply chain for reuse or re-processing into new materials, with as little as possible sent to landfill.

There are numerous strategies that can be employed to minimize the generation of waste associated with the built environment. These include:

- the recovery of waste materials by reintroducing them into the supply chain, either directly back into the process from which they became waste (reuse) or further upstream (recycling)
- designing buildings and infrastructure systems in accordance with standard material dimensions
- controlling the production and disposal of waste through a waste management plan
- improving the efficiencies of manufacturing processes to maximize raw material usage.

The ability to reduce the production of waste in the long term is intrinsically linked to designing for recyclability and reuse. The ability to reuse or recycle materials and products at the end of their initial use minimizes the quantity of materials that is ultimately sent to landfill.

2.2.6 Designing for recyclability

Finishes and services are typically the most often replaced components within buildings. For finishes, this is usually due to normal wear and tear or a desire to change appearance. Services are often replaced due to technological obsolescence or failure. It can be difficult to reuse these elements and so it is essential that finishes and services, as well as other components that have a shorter life than the rest of the building or infrastructure system, are able to be recycled.

However, opportunities for recycling or reusing materials is often limited by the design of the building itself. Regardless of the anticipated life of a building or item of infrastructure, they must be designed with the view that they will ultimately be recycled or reused with minimal additional requirement for resources, for re-processing or making good. Designing buildings and infrastructure so that individual materials can be easily separated can help to facilitate the recycling or reuse processes. Minimizing the time and costs needed to recover materials can often increase the chances that they will be recycled or reused instead of being sent to landfill.

The avoided environmental impacts from the reuse of existing materials and components can vary significantly. The transport distance, degree of making good or re-processing necessary and the quality of materials will all influence the potential environmental benefits of reuse or recycling. For example, materials that are reused on site may require only a small amount of making good and have little need for resource-intensive re-processing. In some cases, the energy required to reuse materials on site can be as little as 5 per cent of the initial energy requirements associated with their production. Materials such as plastic and glass that are taken from a site to be re-processed can require up to 75 per cent of the energy required to manufacture new materials.

When specifying materials for their recyclability or reusability potential, the durability of those materials must also be considered to ensure that their long-term performance and the benefits of long-lasting materials are not being compromised for the benefits offered by the ability to recycle or reuse the materials.

2.2.7 Designing for durability

Durable materials often last longer and require less maintenance. The longer materials last, the less frequent is their replacement and the lower the associated demand for new materials that comes with this. By specifying materials that last as long as possible (with consideration still for their recyclability or reusability when eventually they do come to the end of their service-life), the environmental impacts attributable to these materials can be spread over as long a period as possible. Also, the over-specification of materials should be avoided by considering the intended life of the building or infrastructure. For example, choosing materials that have a 50 year life when they will most likely be replaced within 10 years (in a retail fit-out situation, for example) can require more resources than are necessary in the material's manufacture, when an alternative material may be a more environmentally preferable option.

The durability of individual materials will be determined by not only the properties of the materials themselves but also the function for which they are used and the environment in which they are placed, including their relationship to other materials and their exposure to the weather (Garden 1980). Adequate maintenance can also assist in extending the life of materials, maximizing their resource value and helping to avoid, or at least delay, the demand for additional virgin materials and all of the associated impacts from raw material extraction, processing and the subsequent manufacture of new materials.

As materials become more durable and longer lasting, those that would have otherwise been available as reused or recycled materials for new buildings or the refurbishment of existing buildings and infrastructure will be in decline. Unless large quantities of renewable materials are able to be widely available, resource depletion may actually increase despite our efforts to minimize it through the utilization of more durable, longer-lasting materials. Just as most of the current virgin materials we use in the built environment are not renewable, neither is our second-hand material resource able to be infinitely replenished.

2.2.8 Designing for adaptive reuse

Eventually, most elements within the built environment will come to the end of their useful life. This occurs for a number of reasons, including technological, social, economic, safety, functional and cultural reasons (Langston *et al.* 2008). The ability to continually reuse buildings or infrastructure systems even amidst changing cultural, technological and functional demands is one of the best solutions to reducing the environmental impacts from the built environment. In order to facilitate the reuse of buildings and infrastructure for as long as they are structurally capable of being reused, they should be designed to be adaptable to changing demands and uses. This may involve the use of easily accessible service ducts and flexible internal configurations, for example.

By encouraging and allowing for the adaptation of elements within the built environment throughout their physical life, the value of the resources embodied in these elements can be maximized, minimizing the frequency of demolition and replacement and generating the many environmental benefits that a reduction in waste, raw material demand and new construction can provide. However, maximizing the value of those elements that already exist in our built environment also entails the on-going operation of often highly inefficient buildings and systems. The efficiency of existing systems (such as heating, cooling and lighting) and processes must also be improved to minimize operational resource, energy and water consumption.

2.3 An integrated approach to environmental design

A broad, integrated approach to environmental management and impact minimization can have many flow-on benefits. For example, designing buildings to utilize the off-cuts of materials that may have otherwise been sent to landfill minimizes the potential impacts from their disposal and can also reduce the demand for virgin raw materials. Designing to maximize the durability and life of construction materials reduces the frequency of material replacement and the subsequent demand for raw materials, as well as the accumulation of materials in landfill when they are no longer required.

The built environment must be adaptable to changing expectations, ensuring that short-lived materials are recyclable whilst maximizing the durability and flexibility of those elements that form the main structural components. This integrated approach, which considers the implications of every material, element and system on the environment, is undoubtedly the most realistic approach to making significant improvements to the environmental performance of the built environment.

However, whilst environmental design, with its focus on the design stage, provides some of the greatest opportunities for improving the environmental performance of the built environment, it does not preclude the need or potential for environmental improvements to be achieved during the management and use of the built environment.

2.4 Environmental assessment: an essential component of environmental design

Despite the best intentions to minimize the environmental impacts of the built environment, through applying the above principles throughout its design, often a lack of knowledge or awareness by those involved in the environmental improvement process results in sub-optimal outcomes. Decisions are often informed by inadequate or incorrect information or a misunderstanding of the key issues and their significance. This is not at all surprising considering the vast array of decisions that need to be made and the multitude of choices available. How can designers and those involved in the construction and management of the built environment know what the best choices are?

In order to achieve the ambitious environmental goals necessary to avert further adverse impacts on the environment, comprehensive information is needed to inform the optimal environmental choices during the planning, design and construction of the built environment. *Environmental assessment* provides a means for developing this information and is essential for targeting environmental improvement efforts and verifying environmental claims. The full environmental impacts or benefits associated with the production of a construction material such as steel, for example, cannot be known unless the resources required (raw materials, energy and water) and emissions or waste produced during every stage of the production process, and all of the processes that are required to support it, are quantified. Only then can comparisons be made between competing choices and environmental improvement efforts targeted towards those processes or areas from which the greatest impacts occur.

Despite a practitioner's efforts at reducing impacts through the employment of various environmental strategies, a comprehensive assessment and understanding of these impacts remains vitally important, as it enables those involved to gain a clearer indication of the likely impacts in order to prioritize and implement appropriate improvement strategies. In some cases, impacts associated with some of the strategies that may intuitively be considered to be environmentally beneficial may in fact be greater than those associated with the conventional approach or alternative. Environmental assessment is thus also important in these circumstances to justify certain decisions or to ensure that environmental performance has been improved, as initially intended.

For many decades now people have sought to assess the environmental impacts associated with the built environment in order to identify and minimize them. Numerous tools have been developed to assess these environmental impacts, each with its own focus, benefits and limitations.

2.5 Origins and historical perspective of environmental assessment

Environmental assessment is an approach or process that is used to assess the environmental impacts of any activity. An assessment of environmental impacts involves the quantification of changes to the natural environment resulting from any process or activity. These changes typically result from the consumption of the Earth's resources, the release of pollutants into the environment (such as chemicals and heavy metals) and the disposal of waste. The purpose of any environmental assessment is to inform decision makers of the potential environmental impacts so that actions can be taken to avoid or minimize them (Elliot and Thomas 2009). Environmental assessment in the built environment can be and is used for a range of purposes, including to:

- validate the compliance of building designs with building regulations
- identify the most critical aspects of the environmental performance of existing buildings and infrastructure
- choose between competing or alternative material or component options
- compare design solutions for new buildings and infrastructure and for upgrades to existing buildings and infrastructure
- market the environmental credentials of a building to prospective purchasers or tenants.

The impact of human activities on the environment has been a concern for a number of centuries (Gari 2002). During the modern environmental movement, considered by many to have originated from Rachel Carson's *Silent Spring*, published in 1962 (Carson 1962), concern over the limitations of energy and raw material resources led to a renewed interest in developing methods for quantifying energy and resource use and projecting future energy and resource demands. Numerous studies were conducted in the late 1960s and into the 1970s that attempted to quantify the environmental impacts resulting from a range of manufacturing processes and products. These studies were the precursors to current life cycle assessment techniques. One of the first of these was a study conducted for the Coca-Cola company which compared the raw materials and fuels used and the associated environmental impacts for different beverage containers (Hunt *et al.* 1974).

In the 1970s, a formal process for conducting an environmental impact assessment (EIA) was developed. This was called an Ecobalance or Resource and Environmental Profile Analysis (REPA) (SAIC 2006). Since this time, numerous countries have developed legislation that requires impact assessment to be undertaken for certain projects (US Government 1969; European Union 1985; Australian Government 1999).

With the energy crises of the 1970s came a focus on reducing energy demand, particularly from buildings. This led to the development of *energy analysis* as an approach to the use of environmental impact assessment for monitoring and minimizing resource consumption. These analyses tended to focus on the energy required for heating and cooling of buildings and that consumed during certain manufacturing processes. Due to the abatement of the energy crises and improvements to energy efficiencies as a result of this, broad interest in environmental assessment began to decline in the late 1970s and early 1980s (SAIC 2006). Throughout the 1980s, growing concern about the space needed for solid waste disposal led to a small number of studies being conducted in this area, with a particular focus on packaging materials (Pedersen and Christiansen 1992). However, these studies were often commissioned in order to improve internal production processes and rarely made it into the public domain.

The 1990s saw the beginning of the broader environmental concerns that remain the current focus of environmental improvement efforts today. Much of this renewed interest can be attributed to the Brundtland Report (Brundtland Commission 1987), published by the Brundtland Commission (formerly the World Commission on Environment and Development) in 1987. The need for more complex models of environmental assessment led to the development of a range of guidelines (inter alia Heijungs *et al.* 1992; SETAC 1993; Jensen *et al.* 1997; Todd and Curran 1999) and standards in an attempt to standardize the many different approaches that were being used in the assessment of environmental impacts, which often resulted in conflicting outcomes. At this time, the International Organization for Standardization (ISO) began developing standards for environmental management (the ISO 14000-series) which incorporated the first standards on life cycle assessment (LCA), released in 1997.

2.6 Environmental assessment in the twenty-first century

There is now a renewed interest in improving the environmental performance of the built environment, due largely to the growing awareness of the impacts that it is having on the environment. New standards continue to be developed, such as those by the European Committee for Standardization (under CEN/TC 350) (CEN 2010), and environmental assessment is increasingly being seen as having a central role in improving the sustainability of the built environment. The potential financial benefits of improved environmental performance are also being increasingly acknowledged and demonstrated. This is also one of the driving forces behind the current increased uptake of *green buildings* and the investment in broader efficiency improvements within building and infrastructure construction and material production practices.

Traditionally, environmental assessment has focused on only a limited range of environmental parameters associated with a specific process or activity at any one time. However, unless a detailed and comprehensive analysis is performed, analyzing all of the potential impacts across all life cycle stages of the built environment, how can one be sure that environmental performance is actually being improved through the selection of specific materials, systems and components? With this in mind, more recently the focus of environmental assessment has broadened. An awareness that the built environment results in a much greater range of impacts on the natural environment and human health, than just those associated with energy consumption and waste production, is growing. Many of these impacts are considered to be even more detrimental than those associated with energy consumption alone, particularly with regard to the impacts on human health (such as the release of toxic pollutants into waterways and the atmosphere). Some of the approaches for assessing environmental impacts that are now being used and developed address this much broader range of potential impacts.

2.7 Approaches to environmental assessment

An increasing focus on improving the environmental performance of human activities has led to the rapid expansion in the number of techniques that are now available to assist with identifying and quantifying environmental impacts. These techniques include assessment tools, simulation tools, checklists and guidelines (Arbor 1999; Crawley and Aho 1999; Jönsson 2000), ranging from the very simple to the much more complex, and rely on different levels of data and professional input.

2.7.1 Assessment tools

Assessment tools typically use a database, specifying the inputs from and outputs to the environment related to generic materials and components used in the construction of buildings and other infrastructure systems and products. By selecting from a range of design solutions, the environmental performance of each alternative can be modelled and compared. These tools help to streamline and standardize the environmental assessment process, theoretically saving time and money. Environmental assessment tools are most useful for informing the environmental design or improvement process, but their use does not necessarily lead to a more sustainable built outcome. It is here that professional judgement is essential to identify the best solutions and areas where further opportunities for reducing environmental impacts may be possible. A selection of the tools available for quantifying the environmental performance of buildings is described in more detail in Chapter 4 (Section 4.6). Rating tools, which are used to rate the environmental performance of a building based on particular design solutions, also come under this category. They provide a rating against predefined environmental performance criteria and benchmark performance, such as the quantity of energy required for building heating and cooling. These tools are typically used for demonstrating regulatory compliance and are rarely used to inform environmental improvement beyond the minimum required standards.

2.7.2 Simulation tools

A simulation tool simulates or predicts how a building may perform under certain physical and environmental circumstances in order to identify the potential to improve its environmental performance. One of the most common types of simulation tools is those that focus on predicting energy consumption during building operation, although they are also used to model natural and artificial lighting, air flow and ventilation. Simulation tools often do not convert the environmental loadings or characteristics (such as energy consumption) to an impact on the environment, which is left up to the practitioner to do. However, they can be useful for informing design decisions that may improve environmental performance, such as modifying the design to reduce energy demand, for example.

2.7.3 Checklists and guidelines

Checklists and guidelines provide advice or guidance to help improve environmental performance, often across a broad range of parameters. They do not involve detailed or sophisticated modelling or assessment of the environmental performance of alternative design solutions, but rather help to direct the focus of designers towards the major areas that should be addressed, based on generic knowledge of how buildings and infrastructure can affect the natural environment.

Despite there being a large number available, many of the techniques for environmental assessment do not facilitate the flexible, integrated and comprehensive approach that is needed to maximize the environmental performance of the built environment, nor do many of them consider that impacts often occur across more than one point in time or geographic location. They often rely on a design outcome having already been relatively resolved, in which case major changes – which may actually be necessary to achieve the best environmental outcomes – are often unlikely to occur. In addition, many of the approaches focus on a limited range of environmental parameters or life cycle stages, or may rely on inconsistent or incomplete data and assessment techniques. As a response to the limitations of many of the tools and techniques used for quantifying environmental impacts in the past, the use and development of life cycle assessment (LCA) is now seen to be of critical importance (inter alia Lave *et al.* 1995; Cole 1998; Azapagic and Clift 1999). LCA aims to minimize the limitations of past assessment methods and increase the range of assessment criteria and depth of analysis (Cole 1998). Due to its universality of application, it is considered the only valid method by which to compare the environmental impacts resulting from the alternative building materials, products, components and services that are used within the built environment. Life cycle assessment, including the general framework for its application, is described in detail in Chapter 3.

2.8 Summary

Significant improvement to the design, construction and operation of the built environment is needed to ensure that current standards of living are able to be maintained or improved. At the same time, this needs to be done in a way that sustains or improves the current state of the natural environment. There is a range of opportunities for reducing the environmental impacts of the built environment, including:

- switching from the consumption of and reliance on non-renewable energy and materials to those that are renewable and less polluting
- reducing natural resource consumption through design rationalization and efficiencies in the production of built environment components and during a building's operation

- minimizing the production of waste through more efficient production, reuse and recycling of materials
- considering the durability and recyclability of materials and components
- prolonging the life of buildings and infrastructure systems by ensuring that they can be adapted to changing technological, social and functional expectations
- designing to facilitate reuse or recycling, through the ability to disassemble buildings and infrastructure.

An approach for quantifying the potential environmental impacts from the built environment is essential in order to compare various options, improve existing practice and identify opportunities for environmental improvement that may otherwise have been overlooked. Current techniques for environmental assessment have developed over several decades and interest in environmental improvement has fluctuated over time due to the advent of various environmental concerns. Life cycle assessment currently provides the most comprehensive framework for assisting in the decision-making processes needed to achieve significant environmental improvements within the built environment.

3   Life cycle assessment

There are many types of tools and techniques available for improving and managing environmental performance, each of which provides a unique approach to assessing and potentially identifying strategies for improving the environmental performance of a wide range of products, processes and services. Life cycle assessment (LCA), also known as life cycle analysis or cradle-to-grave analysis, is a more recent addition to these tools, and aims to support an increasing focus towards environmental improvement and management. Life cycle assessment is distinct in that it takes a more holistic approach to environmental assessment, considering a broad range of potential impacts, and as the name suggests, across every stage of a product's life. Another benefit of LCA is that it is based on a chosen functional unit and can factor in many geographic, technological and temporal variations in its assessment.

This chapter provides an overview of life cycle assessment and a description of the various components of a life cycle study, including its various uses and limitations. The various methods of obtaining environmental data are described in detail and the limitations of each of these methods are explained. Particular emphasis is placed on this aspect of an LCA as this phase is the most crucial in ensuring that eventual outcomes of any LCA study are as reliable and useful as possible.

3.1 What is life cycle assessment?

Life cycle assessment is a tool that is used to determine and evaluate the environmental loadings and impacts of a particular product or process, including those effects associated with processes upstream in the supply chain (Curran 1993). LCA is used to analyze these loadings over the entire life cycle of the product or process being studied, from the extraction of raw materials from the ground to final disposal or recycling of the product at the end of its useful life. The environmental impacts across the life cycle of any product or process can be linked to the relevant inputs and outputs of the product system. These inputs and outputs include the raw materials extracted from the earth; energy and water use; emissions to air, land and water; solid wastes; co-products; and other releases, as shown in Figure 3.1.

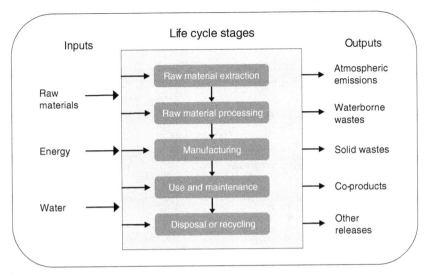

Source: Adapted from Curran 1993.

Figure 3.1 Stages of a product's life cycle, showing inputs and outputs

An LCA does not consider the economic or social aspects of a product, process or decision; however, the same principles applied in an LCA can also be applied to these aspects. A life cycle approach to building economics is often advocated by the proponents of sustainable development, as the typically higher capital cost of this type of development can usually only be justified by the on-going operational cost savings and benefits. Only through a life cycle costing (LCC) approach can these potential benefits and savings be identified. This approach has been detailed extensively by authors such as Langston (2005) and within the International Standard 15686-5 (2008) on life cycle costing.

3.1.1 Life cycle assessment framework

There are four phases to any LCA, as depicted in Figure 3.2. These are defined by the International Standard for Environmental Management (International Standard 14040 2006) as goal and scope definition, inventory analysis, impact assessment and interpretation. These phases are described in detail in Section 3.3.

3.1.2 An iterative approach

LCA involves an iterative approach, where each phase relies on, and can also feed back to inform, the outputs of the previous phase. As a study progresses, a practitioner may identify gaps in data, data reliability issues or significant issues that were previously overlooked. Once these gaps and issues have been identified, particular aspects of the LCA may be re-visited. This iterative approach is important in LCA studies as it provides a level of flexibility and the ability to easily re-visit aspects of the study considered to be deficient in one way or another or that may benefit from the new knowledge gained from the study itself.

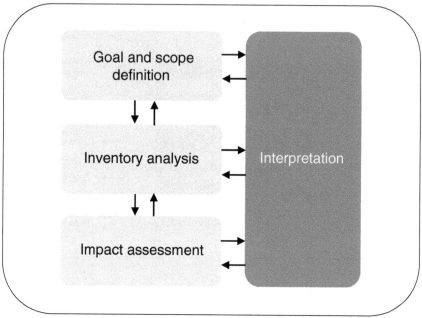

Source: International Standard 14040 2006.

Figure 3.2 Phases of a life cycle assessment

Whilst conducting each phase of an LCA, the practitioner may find it necessary to refine any or all of the previous phases, based on the findings obtained. The findings from one phase may highlight deficiencies in the scope or in the collected data that restrict the ability for achieving the intended project goals. For example, issues with the quantity or quality of data collected during the inventory analysis phase that may limit the ability of the findings to address the study goals may only be identified whilst conducting the impact assessment. The practitioner may then need to redo aspects of the inventory analysis, including collecting more or better quality data, in order to ensure that the project goals are able to be met.

Findings are often used to inform further product development. With product improvement and refinement comes a need to reassess environmental performance to determine whether or not, and to what extent, the environmental goals of the development process have been achieved. The iterative nature of LCA facilitates this process.

3.2 Types of life cycle assessment

Whilst an LCA by its nature generally incorporates an assessment of all aspects of a product's or process's life cycle, the wide range of potential applications (discussed in Section 3.4) for LCA means that the general LCA framework and scope is often adapted to suit individual project goals and needs.

3.2.1 Baseline life cycle assessment

A baseline LCA, better known as a conventional LCA, is used to assess an individual product, system or process, with the ultimate goal of identifying areas of potential improvement in the environmental performance of that product, system or process. This often starts with assessing the environmental impacts associated with a product and using the findings from the LCA to identify the areas where the most significant impacts are occurring. Strategies can then be developed and employed in order to target these key areas.

An example of how a baseline LCA can be used is in the development of new building materials or improving the environmental performance of existing building materials. An LCA, in this context, can be used to identify the inputs of materials, energy, water and other resources, as well as the outputs (pollution, products, co-products and waste) associated with the manufacture of these materials (based on the specific processes, technologies and geographical locations involved). The findings of this study, potentially detailing the quantity of energy, water and raw material consumption and waste associated with the manufacturing process, can then be used to identify which process or stage of manufacture or specific raw material or other resource input contributes the greatest to the overall environmental impact associated with the manufacture of the building material. Armed with this information, the product designers are then able to either substitute new processes, materials and resource requirements for existing ones or focus on improving the efficiency of these existing processes.

The findings from a baseline LCA are often used for internal purposes as this type of LCA particularly lends itself to product development and improvement. However, the findings are also often used to market particular environmental strengths or qualities of a product to meet the demands of an increasingly environmentally conscious market.

3.2.2 Comparative life cycle assessment

A comparative LCA, as the name suggests, is used to compare the environmental impacts of two or more products or processes that perform the same function. This can be useful where the intended purpose is to select the alternative with the lowest environmental impact. With any comparative LCA, it is essential that a comparable functional unit is selected so that the alternatives are compared on a common basis. For example, comparing the environmental impacts of two light globes with varying light output would not be an entirely meaningful comparison to make. In this case, the quantity and quality of light for the particular task is the most crucial factor in defining the function of the globes. Such a comparison would need to be based on a specified quantity of light output for the findings to be relevant. A greater number of globes of lower output may be required to provide a similar function (lighting a specified area to a certain level of illuminance) to a globe of greater output.

Product development often begins with a baseline LCA in order to determine the most significant environmental impacts. As alternative products or solutions are developed in an attempt to reduce these impacts, it may be necessary to perform a comparative LCA that compares all alternatives to determine whether environmental improvements are being made. This type of assessment is useful for decision makers where more than one option is available.

3.2.3 Streamlined life cycle assessment

Quite often, the goals of an LCA study are to assess only a limited range of environmental impacts or life cycle stages associated with a product or process. Previous work may have identified significant impacts resulting from a particular input (such as energy) or life cycle stage (such as manufacture). For example, where alternative manufacturing processes are being investigated for a single product, it may be acceptable to study only this stage of the product's life cycle. This type of LCA, where the scope is limited in depth, breadth or detail, is known as a streamlined LCA. The goals and scope, identified at the beginning of the study, will inform the appropriateness of using a streamlined approach. Caution must be exercised in using this approach as a much more detailed understanding of a product is often necessary in order to limit the scope of the study in this way. There must be a good reason why the scope has been limited, which should be explained in the scope definition phase.

It has been shown that the greatest impacts associated with buildings are often related to the consumption of energy (Junnila and Horvath 2003); therefore the focus of much of the environmental performance improvement in the construction industry has been on the energy efficiency of the various stages of a building's life, from construction to use and eventual disposal. For this reason, a streamlined approach that considers only the energy-related environmental impacts is quite common for LCA studies of buildings. This type of study is commonly referred to as a life cycle energy analysis.

3.3 The four phases of life cycle assessment

3.3.1 Goal and scope definition

The first phase of an LCA, the definition of the goals and scope of the study, involves determining the reasons for carrying out the study and the intended audience. The goal and scope definition phase is important in determining the direction and extent to which the LCA will be conducted. This process will determine the approach taken in analyzing the inventory and the detail to which this is performed. The scope of an LCA includes defining which life cycle stages are to be considered. Depending on the product being assessed, these typically cover the extraction of raw materials from the earth, raw material processing, material or product manufacture, operation or use, maintenance and repair and disposal or recycling (International Standard 14040 2006). This phase of an LCA is the most important, as obtaining the correct results for the required purpose is dependent on sufficiently defining the breadth, depth and detail of the study (International Standard 14040 2006).

3.3.1.1 Goals

The defined goal of the LCA study will guide the assessment process to ensure that the most useful results are obtained. Whilst the primary goal of an LCA will generally be to choose the product or process that has the least effect on the environment and human health or to guide the development of new products or processes, the specific goals of an LCA study must reflect the intended use of the findings and respond to the reasons for carrying out the study. The goals must also state the intended application of the study findings, to whom the results of the study are intended to be communicated (intended audience) and whether results are intended to be used for comparative assertions which may be released to the public.

3.3.1.2 Scope

Defining the scope of an LCA study involves determining which life cycle stages are to be considered and what the boundaries for the product system are, noting any assumptions or limitations to the study and detailing which inputs, outputs and impacts are included and which are excluded, and why. In order to define the scope of the study, International Standard 14040 (2006) recommends a broad range of elements that must be considered and clearly described. These items include:

- the product system to be studied
- the functions of the product system/s
- functional unit
- data requirements
- initial data quality requirements
- allocation procedures
- impact categories selected and methodology of impact assessment and interpretation to be used
- type of critical review, if any
- type and format of report required for the study.

The scope of the study should be defined in a way that ensures that the outcomes of the study are sufficient to address the intended goal and eventual application of the study findings. Just as for the definition of study goals, the scope of an LCA study will be influenced by the intended use of the findings, including the level of breadth and detail needed to address the study goals and the level of reliability, detail and confidence required of these findings. These may be influenced by external bodies, the need for consistency or comparability between independent studies or the need or wish to comply with certain standards, such as International Standard 14040 (see Section 3.5).

As an LCA study can involve a highly iterative process, the scope, as well as other aspects of the study, may require modification as the study progresses in order to meet the original goals and intentions of the study. The need to modify aspects of the scope will become evident as the study progresses and findings are evaluated. This modification may be as a result of unforeseen limitations or constraints or due to additional information becoming available.

3.3.1.3 Functional unit

The functional unit defines the performance characteristics of the product or process being studied in order to fulfil a particular function. It is used as a unit of reference, particularly to enable comparative assertions to be made about competing products. The main purpose of defining a functional unit is to ensure that the quantified inputs and outputs are related to a particular function. This is essential for ensuring the comparability of LCA results, which is only possible if the comparisons of products or processes are performed on a common basis.

The chosen functional unit will depend on the particular function or application of the product or process being studied. For example, the comparison of floor covering materials on a cubic metre basis may be unacceptable, as the materials may have different thicknesses and densities. Analyzing them on a square metre basis for use over a certain period of time is much more appropriate, as this matches more closely their intended use. By using a common functional unit, the environmental performance of two alternative floor covering materials, such as carpet and ceramic tiles, can be compared. For each alternative it is possible to quantify the inputs and outputs of each product system. In this case, this would include the materials, other resources, energy and water required during the production, use and eventual disposal stages for these two options, as well as the resultant waste and emissions from these processes.

The actual size of the functional unit is quite arbitrary; however, it may simplify the impact assessment and interpretation phases of the LCA if the size is easily understandable or in a common unit of measurement. An assessment of the impacts for various floor covering materials may be much more useful on a 1 m^2 basis than any other size, particularly when the total impacts for the floor coverings in a whole building need to be determined. All inputs and outputs are relative to the functional unit, as are the resultant environmental impacts. An increase in the quantity or size of the functional unit would typically result in an equivalent increase in the associated inputs, outputs and impacts.

3.3.1.4 System boundaries

A product system is made up of many individual inputs, outputs and processes. The system boundary defines which of these inputs, outputs and processes are to be included in an LCA study. Inputs often include raw materials, ancillary materials, intermediate materials or products, energy, water and other resources from nature. Outputs include such things as waste and emissions to the air, water and land, as well as the final and any intermediate products that are produced (Figure 3.3). It is necessary to define the level of detail to which each of these aspects will be quantified. Resource requirements for certain processes may be included whilst those for other supporting processes may be excluded for varying reasons. Many of these exclusions occur without the knowledge of the practitioner due to their lack of understanding of the product system or choice of data or assessment method.

Often the limiting factor in defining the system boundary is the ability to gain sufficient (or any) data for particular inputs, outputs or processes. These limitations are often based on cost and time constraints or more often on a lack of understanding of all of the likely inputs, outputs and processes required to support any product system. For example, many studies exclude the quantification of inputs and outputs associated with service-related processes (e.g. financial, communication and marketing services) that are necessary to support many manufacturing-type processes. This issue is covered in more detail in Section 3.6.2.

The International Standard for 14040 (2006: 12) permits the omission of particular inputs, outputs or processes if this omission does not significantly change the overall conclusions of the study. The issue here is that a system boundary may be drawn that excludes such processes without sufficient knowledge of their significance. It can be very difficult, prior to any data being collected, to know whether certain inputs, outputs or processes will or will not significantly change the overall conclusion of a study. Whilst the iterative nature of an LCA does allow for the inclusion of additional inputs and outputs during the course of a study if they are identified or found to be significant, the practitioner should not preclude this from being possible by limiting the system boundary too early in the study.

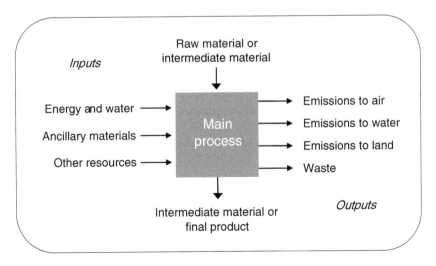

Figure 3.3 Inputs and outputs for a main process

3.3.1.5 Data quality and scope

The quality of the data used for an LCA study can have considerable implications for the ability to use the study findings for a particular purpose and the reliability that can be given to these findings and any assertions based on them. When defining the requirements for data quality, it is important that the intended goals and scope of the study are kept in mind to ensure that these are able to be met. Some of the considerations that should be addressed when defining the quality of data required include the age of the data, the geographic location from which the data is to be collected, the variability of the data values, the degree of completeness, how representative the data is of the particular processes and technologies, the source of the data and the certainty that can be attributed to it (International Standard 14044 2006).

Another factor that may be considered during this scoping phase is the extent to which collected data should comply with specific geographic, technological or temporal characteristics of the product or process under analysis. Significant data variations can occur across each of these aspects, even for similar products or processes. In some cases, it may be necessary to use data that varies in one or more of these areas from those geographic, technological or temporal characteristics specific to the product or process being analyzed. The acceptability of using this data should be considered at this stage, particularly with regard to the effect that this may have on the study findings and subsequent conclusions and recommendations (Weidema 1997).

3.3.2 Life cycle inventory analysis

Inventory analysis, the second phase of an LCA, involves the collection of data and calculations in order to quantify the inputs and outputs to the product system over its entire life cycle. The inputs may include energy, water and natural resources, whilst outputs may include emissions to air, water and land (as shown in Figure 3.3). The extent of this analysis, and the two subsequent phases of the LCA, depends on the scope and goals defined in the previous phase. The inventory analysis is one of the most time and cost-consuming processes in an LCA (Suh and Huppes 2002). If the scope of the study is not adequately defined, excessive time may be wasted obtaining and analyzing data that is beyond that required for the intended purpose of the study. Section 3.6 deals with this and other problems associated with the inventory analysis phase of an LCA.

There are several steps involved in the collection of data for the inventory analysis. International Standard 14044 (2006) outlines these as:

- drawing specific process flow diagrams, outlining all unit processes to be modelled, including direct and indirect processes
- describing each unit process in detail and listing data categories associated with each unit process
- developing a list that specifies a unit of measurement

- describing data collection techniques and calculating techniques for each data category.

These steps are required to ensure that there is thorough knowledge of each unit process in order to avoid double counting or gaps. Problems associated with quality and completeness of data collection are dealt with in Section 3.6.4.

3.3.2.1 Data types

There is a wide range of data types that may need to be collected for an LCA study. The type of data needed will vary depending on the scope of the study, the product or process being analyzed, the inputs and outputs being assessed and the impact categories being considered. Types of data that may be required include energy or water consumption figures associated with material extraction or manufacturing processes and product use; quantities of raw materials consumed during the production of materials or products; or quantities of waste produced and emissions released from material extraction, manufacturing or disposal.

3.3.2.2 Quantifying inputs and outputs

The main purpose of this task is to identify the significant inputs and outputs of the processes or product system. There is a variety of methods that can and have been used for quantifying these inputs and outputs, each having their own benefits and limitations – process analysis, input-output analysis and hybrid analysis.

PROCESS ANALYSIS

A process analysis uses a combination of process, product and location-specific data to calculate the environmental loadings and related impacts of a product system. This data is actual measured data specific to a particular product or process. A process flow diagram can be drawn outlining all of the individual processes to be modelled within the product system. This helps in the identification of those processes for which data will need to be collected. Figure 3.4 shows an example of a process flow diagram for the production of concrete.

Once the data requirements have been identified the data can then be sourced. For an LCA study of a particular product or process that is produced or occurs in a known location or under known conditions, process data may be sourced from internal company records. These records may include energy or water bills or reports that may have been prepared to comply with internal procedures or third-party regulatory requirements, such as those reporting the emission of pollutants and other substances to bodies that track national pollution levels (such as the UK National Atmospheric Emissions Inventory (NAEI), the US Toxic Release Inventory (TRI) and the European Pollutant Emission Register (EPER)).

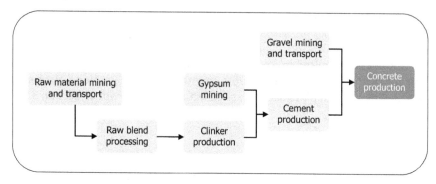

Figure 3.4 Example of a process flow diagram for concrete production

The accessibility to process data can have a significant impact on the time and costs involved in compiling a life cycle inventory (LCI) using a process analysis approach. Much of the data required may be difficult to obtain or may not exist in the format required. In this case, it may be necessary to either spend time further analyzing the specific production processes to collect the necessary data or to use existing generic data collected for similar products or processes. Whilst the latter of these alternative approaches to sourcing data may be much less time-consuming, issues may arise with the reliability and appropriateness of such data when analyzing a specific product or process. This type of data, often an aggregate of data across a number of manufacturers or suppliers, is typically provided in many of the commercial LCA software tools and databases (such as the *Ecoinvent database*, developed by the Swiss Centre for Life Cycle Inventories). Generic process data very rarely provides a completely accurate indication of the inputs, outputs and impacts associated with any specific product or process. Environmental impacts from any product system will vary due to any number of factors, including source, type and quality of raw materials; fuel mixes; manufacturing processes and technologies; use of pollution control strategies; transportation distances; and the treatment of waste. These impacts will very rarely be identical across different manufacturers or suppliers due to these factors. As these factors will also change over time, the age of the data will also influence how accurate it might be for a particular product or process at a given point in time. For these reasons, using generic process data, whilst convenient, can also be highly problematic. Where the use of this data is considered acceptable for a particular study, any impact that this may have on the eventual study findings and recommendations should be clearly understood, possibly through conducting a sensitivity analysis.

Process analysis is generally seen to be more accurate and relevant to the product or process being analyzed; but on the other hand, it can be labour and time intensive. Moreover, it suffers from a systematic incompleteness, which is due to the representation of the product system by a finite boundary, and the omission of contributions outside this boundary. This means that an

LCI based on a process analysis approach does not usually cover the product system to a sufficient degree. The magnitude of this *truncation error* depends on the type of product or activity, but can be around 50 per cent for moderately complex products (see also Norris 2001) and up to 87 per cent for more complex products (see for example Born 1996; Hendrickson *et al.* 1998; Lenzen and Dey 2000; Crawford 2005, 2008). For very energy- or resource-intensive manufacturing processes, such as metals production, the system boundary issue is reduced, due to the much higher significance of the main process. Treloar (1997) and Lenzen (2001) have shown that within a conventional process analysis, this error is not usually reducible to an acceptable level by further extending the system boundary because of the complexity of inputs and outputs that would have to be investigated.

INPUT-OUTPUT ANALYSIS

Input-output analysis is a top-down economic technique, which uses matrices of sector-based monetary transactions (input-output tables) describing complex interdependencies of industries in order to trace resource requirements and pollutant releases throughout a whole economy. These input-output tables describe the inputs required by each economic sector from each and every other sector in order to produce a certain quantity of output. They are most commonly available in economic units. Using this matrix approach, it is possible to predict the effect of changes in one industry (such as increased or decreased output) on others. It is often used to show how a particular activity, such as a government policy or an event like the Olympics, provides direct and indirect benefits to the economy, in terms of gross domestic product or jobs. Input-output tables are produced by a significant number of countries worldwide, including Australia, the United States, Japan, United Kingdom, China, India and many countries throughout Europe. The complexity of these tables can differ considerably by country, particularly in relation to the number of economic sectors into which the entire country's commodity output is aggregated. This can range from several dozen sectors up to hundreds. Table 3.1 considers a simplified input-output matrix (also known as a *total requirements coefficient matrix*) depicting the value of outputs from the *Iron and steel*, *Sawmill products* and *Plastic products* sectors purchased by the *Residential building*, *Other construction* and *Prefabricated buildings* sectors in order to produce a certain value of outputs from each of these sectors.

This simple input-output matrix shows that per unit of output from the *Residential building* sector, 0.15 units worth of output from the *Iron and steel* sector (15 divided by 100), 0.20 units worth of output from the *Sawmill products* sector (20 divided by 100), 0.05 units worth of output from the *Plastic products* sector (5 divided by 100) and 0.60 units worth of labour (60 divided by 100) are required. From this, it is possible to measure the impact of any increase in output from a sector, in terms of the increased requirement for inputs from the supporting sectors.

Table 3.1 Example input-output table showing total value of inputs per unit of output
for construction sectors

Selling sector	Purchasing sector		
	Residential building	Other construction	Prefabricated buildings
Iron and steel	15	30	20
Sawmill products	20	10	15
Plastic products	5	10	5
Labour	60	80	40
Total output	100	130	80

Generally, input-output analysis covers completely the economic system
defined by the national or regional statistics. The *total requirements coefficient
matrix* (an innovation for which Wassily Leontief won the Nobel prize) treats
the whole economy as a system and any number of inputs from other sectors
are included, resulting in an almost limitless number of potential transactions
upstream through the supply chain. Generalized input-output frameworks
have been applied extensively to environmental analysis since the late 1960s
(see for example Isard *et al.* 1968; Isard and Romanoff 1967; Leontief and Ford
1970; Bicknell *et al.* 1998). An introduction to the input-output method and its
application to environmental problems can be found in studies by Leontief
and Ford (1970), Proops (1977), Duchin (1992) and Dixon (1996). In order to
use an input-output table to identify the inputs, outputs and impacts
associated with any commodity, it must first be integrated with sector-based
environmental data.

Almost any type of environment-related data can be integrated into an
input-output table, including energy, water and raw material consumption, as
well as production of waste and pollutant emissions, such as carbon dioxide.
To enable the integration process to occur, the environmental data must be in
the correct format, giving values of inputs of resources or outputs of pollutants
or waste per economic sector (for example, gigajoules of energy or tonnes of
carbon dioxide emissions). By knowing the total resource requirements or
emissions associated with a particular sector as well as the total monetary
value of all outputs from that sector, it is then possible to determine the
quantity of resources required or emissions released per monetary unit of
output for each sector. This is done by dividing the total resource requirements
or emissions by the total monetary value of sector outputs, which results in a
sector-level *total requirements coefficient* in units per monetary unit of output
(e.g. tonnes of CO_2 per dollar).

By integrating environmental data into the input-output table, it is then
possible to determine the quantity of resources or emissions associated with a
particular value's worth of output from any sector of an economy, including
the *indirect* (or upstream) inputs of goods and services required to support the
production or supply of that output. One of the most common forms of
environmental input-output table is one based on energy flows or requirements

(for example Carnegie Mellon University 2002; Lenzen and Lundie 2002). In this case, national energy data is integrated with the input-output table by multiplying the direct requirements coefficients describing sales by energy supply sectors and the appropriate national average energy tariffs, to give direct energy requirements of individual sectors (Equation 3.1). This process is explained in more detail by Treloar (1998: 25). The modified input-output table forms the input-output model that can then be used to quantify the *total energy requirements* associated with the production of any product or supply of any service within an economy, based on its monetary value.

$$DER_n = \sum_{e=1}^{E} D_{en} \times tariff_e$$

Equation 3.1

Where DER_n = the direct energy requirement of the target sector, n; E = the number of energy supply sectors, e; D_{en} = the direct requirements coefficient for each e into n; and $tariff_e$ = energy tariffs for each e, in units of energy per dollar.

The single biggest problem with almost all previous applications of input-output analysis is that it is used as a black box. The *total requirements coefficients* include the *direct requirements* and the *indirect* (or upstream) *requirements* all added together, for any sector. It can be extremely time-consuming to manually trace the *indirect requirements* for a sector. The *direct requirements* are there for everyone to see, and the sum of the *indirect requirements* can be deduced by subtracting the *direct requirements* from the *total requirements*; however, modelling individual *indirect requirements* (such as the energy required to mine the iron ore needed to produce steel) is much more complex due to the sheer number of them.

Let us assume that an input-output model is disaggregated into 106 industry sectors. One transaction upstream of a main process (such as building construction) involves potentially 106 inputs at the sector level. Assuming only about 8 per cent of the 106 x 106 matrix (giving the inputs from each sector into each and every other sector) of direct requirements are zero values, this reduces on average the number of potential inputs to 98. At one transaction upstream, we might have concrete slurry used in building construction. The potential number of non-zero pathways increases exponentially at an alarming rate with each further transaction upstream. At two transactions upstream, the number of potential non-zero pathways is 98^2. A transaction at this stage might be cement used in the concrete. At three transactions upstream, we have 98^3 additional processes at the sector level. This might include recovery of raw materials such as limestone used for making cement. At four transactions upstream, we have 98^4 potential transactions, which might include the provision of capital equipment to mine the limestone, amortised over its life (i.e. just the share of that equipment's value used to mine that limestone). At five transactions upstream, we have 98^5 potential transactions, which might include the manufacture of sheet steel to make the machinery used for mining the limestone. These numbers are starting to get beyond our ability to comprehend them, with a total number of potentially non-zero pathways at the sector level, at five stages upstream, according to these simplified calculations, of 9,132,395,678.

The system boundary of input-output analysis is economic, such that if a sector pays for any product or service, the inputs to that product or service are counted. It is this power that is the main benefit of input-output analysis. However, the same features that give the technique this power are also its greatest source of error. Aggregation of commodities and establishments into sectors reduces the relevance of the results to any particular product or region. The economic system boundary that gives input-output analysis its depth and breadth results in errors when these financial flows are attributed to physical quantities of materials or environmental loadings. Having said that, the results of input-output models are often quite close to the results from process analyses, though are often higher due to the greater system boundary completeness.

At best, a detailed process analysis may be capable of quantifying the inputs or outputs of a product system to two or three stages upstream of the main process (Stage Zero). As Figure 3.5 shows, a significant quantity of these inputs or outputs can occur further upstream of this. For example, on average for the Australian economy, represented by 106 sectors (based on 1996-97 input-output tables (ABS 2001a)), 28 per cent of the energy requirements associated with the production or supply of goods and services occurs upstream of Stage Three (three transactions upstream of the main process). This highlights the importance of input-output analysis, as tracing entire supply chains using a process analysis approach is virtually impossible (Treloar 2007).

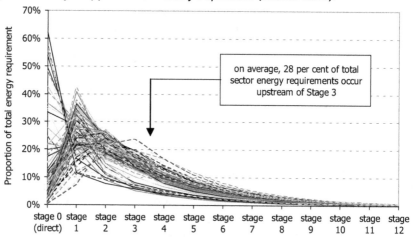

Figure 3.5 Total energy requirements of Australian industry sectors by stage, based on 1996-97 input-output tables

HYBRID ANALYSIS

Due to the quite specific limitations of process analysis and input-output analysis, some have sought to combine the best of both worlds. A number of researchers have suggested and demonstrated a number of hybrid LCI approaches (inter alia Bullard *et al.* 1978; Moskowitz and Rowe 1985; Lave *et al.* 1995; Treloar 1997; Hendrickson *et al.* 1998; Joshi 1999; Suh 2002; Strømman *et al.* 2009).

In a tiered hybrid LCI (Suh and Huppes 2002), the direct and downstream requirements (for manufacture, use and end-of-life) and some important lower-order upstream requirements of the functional unit are examined in a detailed process analysis, while remaining higher-order requirements (for materials extraction and manufacturing) are covered by input-output analysis. In this way, the advantages of both methods – that is, completeness and specificity – are combined. Moreover, the selection of a boundary for the production system becomes obsolete. But this technique still uses input-output analysis as a black box, relying on the consultant to decide which processes are important and require analysis. This can only resolve the upstream truncation error for items that the user decides are relevant. If the user chooses to start the input-output analysis at the highest point in the supply chain (for example, the construction of a building), all opportunities are lost for integrating potentially available and more accurate process data. As the supply chain is disaggregated to allow the integration of process data, the potential exists for sideways (e.g. minor goods and services) and upstream (higher order processes required to support a lower order process) truncation errors as physical flows are almost invariably favoured over the economic system boundary upon which input-output analysis derives its power, as discussed above and as shown diagrammatically in Figure 3.6.

The hybrid LCI model developed by Treloar (1997; 2007) does away with many of these problems by starting with a disaggregated input-output model, in to which available process data is integrated. This avoids the possibility for

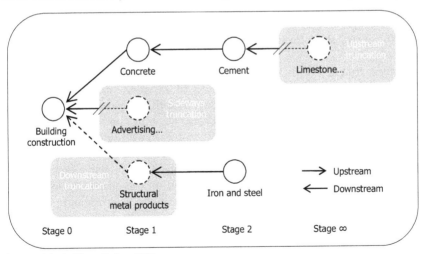

Source: Adapted from Treloar 2007.

Figure 3.6 Upstream, downstream and sideways truncation errors in the building construction system boundary

the truncation errors discussed above. Initially, the input-output model can be used to scope a product to identify the most significant inputs or outputs. It is then possible to target efforts towards reducing the environmental impact of particular processes. The use of this hybrid approach for assessing the resources and emissions attributable to the built environment removes some of the errors associated with previous studies, using process-based methods of assessment. This hybrid LCI approach (known as input-output-based hybrid analysis) is described in more detail in Section 4.4.2.

3.3.3 Life cycle impact assessment

Life cycle impact assessment, or LCIA, the third phase of an LCA, translates the results of the LCI analysis into numerical indicators for specific categories, which reflect the environmental loading of the system or product (Evans and Ross 1998). The purpose of the LCIA is to evaluate the magnitude and significance of potential environmental impacts of a product system based on the findings from an LCI analysis. The LCI findings, which specify quantities of resources, energy, water, waste and emissions associated with a particular product system, are translated into the impact that these environmental loadings may have on the environment. Whilst an LCI will provide useful information for decision making, its usefulness can often be limited by the disconnection between these environmental loadings and their impact on the environment. For example, the environmental impacts associated with the consumption of energy required to manufacture a particular product may differ significantly depending on a number of factors, including:

- type and source of fuel used
- greenhouse emissions intensity of the fuel used
- energy generation processes and technologies
- efficiencies of production and consumption.

Assuming that the consumption of a certain quantity of energy results in similar environmental impacts regardless of variations in the factors listed above would be erroneous. For this reason, quantified environmental loadings, such as energy consumed or waste produced, must be translated into an impact that these loadings have on the natural environment. Similarly, the potential impacts resulting from different outputs, such as carbon dioxide and methane emissions, may vary considerably, even for similar quantities. The LCIA attempts to clarify which is worse, based on the quantified inputs and outputs and their potential hazardous impact on the environment.

The LCIA phase involves the following steps:

- selection and definition of impact categories
- classification
- characterization
- normalization

- grouping
- weighting
- data quality analysis.

3.3.3.1 Selection and definition of impact categories

The first step involved in an LCIA is to define the impact categories associated with the environmental loadings that have been quantified in the LCI. The choice of impacts that are considered and their level of detail are determined during the goal and scope definition phase of the study. Some of the most common impact categories used for an LCIA are listed in Table 3.2.

The inventory analysis, or LCI, results can then be inserted into these categories, based on meaningful value judgements. Again, the level of detail and depth of this step is defined in the scope and goal definition phase of the LCA. Here, it can again be seen that any decisions made during the initial phase of an LCA will influence and have a flow-on effect throughout the entire study. That being said, the iterative nature of an LCA allows for additional impact categories to be considered should they be identified as important to the assessment. This may come about by the discovery of additional inputs or outputs to the product system and/or an impact that these may have on the environment or human health that had not been clearly identified during the scoping phase of the study.

Table 3.2 Common impact categories used in a life cycle assessment

Impact category	Definition
Global warming	Increase in the Earth's average temperature
Depletion of minerals and fossil fuels	Consumption of non-renewable energy or material resources
Photochemical oxidation (smog)	Emission of substances (VOCs, nitrogen oxides) to air
Human toxicity	Human exposure to an increased concentration of toxic substances in the environment
Ozone depletion	Increase of stratospheric ozone breakdown
Eutrophication	Increased concentration of chemical nutrients in water and on land
Water use	Consumption of water
Land use	Modification of land for various uses
Acidification	Emission of acidifying substances to air and water
Ecotoxicity	Emission of organic substances and chemicals to air, water and land

3.3.3.2 Classification

The classification process involves assigning LCI results to the selected impact categories. For example, the release of a range of greenhouse gases may be classified into the *global warming* category, as the release of these greenhouse gases impacts on global warming. Likewise, the release of phosphates and nitrates into the environment can result in an unnatural concentration of these nutrients in an ecosystem and, as such, they are often classified as a *eutrophication* impact. Often, certain inputs or outputs of a product system may contribute to more than one environmental impact and must be assigned to multiple impact categories either on a proportional basis or wholly to each impact category to which they contribute. For example, the consumption of energy could affect both global warming, through the release of greenhouse gases from combustion, and the depletion of natural resources, depending on the source and type of energy used. Additional examples of the classification of impacts are shown in Table 3.3.

Table 3.3 Impact classification for some common impact categories

Impact category	Example LCI data	Characterization factor
Global warming	Carbon dioxide (CO_2) Nitrogen dioxide (NO_2) Methane (CH_4) Other greenhouse gases	Global warming potential (carbon dioxide (CO_2) equivalents)
Resource depletion	Quantity of minerals used Quantity of fossil fuels used	Resource depletion potential (ratio of resources used to quantity of resources left in reserve)
Land use	Quantity of landfilled waste Other land modifications	Land availability (volume of waste)
Eutrophication	Phosphate (PO_4) Nitrogen oxide (NO) Nitrogen dioxide (NO_2) Nitrates	Eutrophication potential (phosphate (PO_4) equivalents)
Acidification	Sulfur oxides (SOx) Nitrogen oxides (NOx) Hydrochloric acid (HCL) Ammonia (NH_4)	Acidification potential (hydrogen (H+) ion equivalents)
Ozone depletion	Chlorofluorocarbons (CFCs) Hydrochlorofluorocarbons (HCFCs) Halons	Ozone depleting potential (trichlorofluoromethane (CFC-11) equivalents)

Source: Based on SAIC 2006.

3.3.3.3 Characterization

The characterization process involves calculating the category indicator results. The results from the LCI are converted to common units based on conversion or characterization factors that indicate particular impacts to human or environmental health. By translating the different LCI results into directly comparable units, they can then be directly compared within each impact category.

For example, greenhouse gases are often expressed in CO_2 equivalents. LCI results detailing the quantity of different greenhouse gases emitted for a product system can be multiplied by specific CO_2 equivalents or characterization factors to provide an overall indication of global warming potential (see example in Section 4.5.3 of Chapter 4). This way, different quantities of different substances can be compared equally to determine the impact that each one has on global warming. Some characterization factors are shown in Table 3.3 for some of the most common impact categories.

3.3.3.4 Normalization, grouping and weighting

After the LCI results have been characterized, they may then need to be expressed in a way that allows for direct comparison across impact categories. This process, known as normalization, divides the indicator results by a selected reference value. These may be compared on a per capita or per unit basis or compared to the highest value among all options. Normalization may provide a clearer indication of the significance of particular indicator results.

Grouping provides a way of sorting impact categories into specific areas. Impact categories may be sorted by their particular characteristics (such as inputs and outputs or by location) or by a ranking of their significance. This grouping of impacts can be useful for the interpretation of findings.

The weighting process involves assigning weights or numerical values to the different impact categories assessed based on their perceived importance. This process must reflect the original goals of the study and the values of the relevant stakeholders. For example, the consumption of water and the resultant depletion of water reserves may be of great concern and of high importance in locations where it is in limited supply but not so much in locations where it is in plentiful supply. Weighting factors allow the importance of different impact categories, such as global warming and eutrophication, to be compared. However, ranking the importance of one potential impact over another is a highly subjective process and relies on value judgements about how significant the potential impacts from each category are compared to those from another. For example, is global warming and its potential impacts on human and environmental health a more important concern than the pollution of our water supplies? Different individuals or organizations will have different priorities and preferences and therefore it is highly likely that they may each reach different findings based on the same indicator results.

3.3.3.5 Data quality analysis

The final stage of the LCIA phase involves verifying the accuracy of the results obtained. The findings from the LCIA stage are only as accurate, reliable and complete as the data sources and collection methods allow. Incomplete or unreliable data will inevitably lead to an inaccurate representation of potential environmental impacts. As each phase of an LCA is dependent on and influenced by each previous phase, it is extremely important that each of these phases is conducted with the intended application of the findings in mind at all times. It is also important that the potential uncertainty and sensitivity of the findings from the LCIA are known. This can be used to inform the level of confidence that can be put into any claims made on the basis of these findings. Uncertainty analysis can be used to determine the level of uncertainty associated with particular data and assumptions and how these might affect the reliability of the LCIA results. Sensitivity analysis is used to determine how any changes in the data or assumptions might affect the results of the LCIA.

As the LCIA phase also involves reviewing the goal and scope of the LCA to ensure that the study objectives have been met, these analyses may also guide the iterative LCA process and result in revision of the other LCA phases. The iterative approach to LCA means that if alterations to the goals or scope of the study are necessary, based on the findings of the LCIA, then aspects of the study can be re-visited to address these changes.

3.3.4 Interpretation

The fourth and final phase involved in an LCA is the interpretation of findings. This phase involves combining the results from both the LCI and the LCIA (Phases Two and Three) in order to determine the most important inputs, outputs and potential environmental impacts of any product system (International Standard 14043 2000). The information obtained from these previous phases is identified, qualified, checked and evaluated. Conclusions can then be drawn and recommendations made, with particular emphasis on the identification of areas for improvement. This step is commonly referred to as *improvement analysis* for this reason.

The main steps involved in this phase include:

- identifying significant issues
- evaluating results
- drawing conclusions
- explaining limitations
- providing recommendations based on the findings of the preceding phases of the study.

3.3.4.1 Identification of significant issues

The information obtained from the LCI and LCIA phases is reviewed to identify the data that has the most influence on the LCA results. This step guides the evaluation of results, so that effort is placed mostly on those impacts that are found to be most significant or represent the greatest contribution, or on the data and assumptions that have the most significant impact on the study findings. The issues that have the most significant impact on the study findings can range from specific inputs or outputs (such as energy use or emissions), impact categories (such as global warming or resource depletion) or particular life cycle stages or processes (such as raw materials extraction or manufacture).

3.3.4.2 Evaluation of results – completeness, consistency and sensitivity

The evaluation step involves determining the level of confidence and reliability that can be attributed to the study findings. The completeness of the study must be assessed to ensure that all relevant inputs, outputs and impacts have been considered. This ensures that all of the data and information needed for the interpretation of the findings is available and complete and that it reflects the intended goals and scope of the study. If information or data is found to be missing then a reliable assessment cannot be performed and it may be necessary to re-address certain aspects of the study in order to fill these gaps.

A consistency check should be conducted to ensure that assumptions, methods, allocation of impacts, data collection and system boundaries are consistent across the entire study, especially where alternative options are being assessed. Inconsistencies within a study may occur due to the data sources used, the geographical or technological coverage of the data, the accuracy and age of the data or the methods used to collect or assess the data and the respective environmental impacts.

A sensitivity check is performed to assess the reliability of the final results and how they may be affected by the data, methods or assumptions used. By testing the potential variations in the methods and data used and the assumptions made, the influence of these choices on the final results can then be ascertained. The results obtained from the study are compared to the results that would be obtained if certain data, assumptions or other study parameters were altered. The sensitivity is then expressed as the possible deviation of results and significant changes can be identified (International Standard 14044 2006). Typically those issues found to be most significant are given most priority when assessing the sensitivity of results. The intended use of the results will direct the level of reliability that is considered acceptable and determine whether or not more extensive analysis may be required. A sensitivity analysis is demonstrated in more detail in Section 4.5.6 of Chapter 4.

3.3.4.3. Conclusions, limitations and recommendations

The final stage of the interpretation phase involves drawing overall conclusions from the study, identifying potential limitations and making recommendations based on the study findings. The overall conclusions are drawn by considering the various factors, such as the most significant issues, the completeness and consistency of methods and results, the intended goals of the study and any limitations. The limitations may be related to the way in which data has been collected, analyzed and interpreted. Recommendations related to the intended application of the findings can then be made that best inform the intended audience and decision makers.

The major recommendation of a comparative LCA study will generally be related to which product or process has the least overall impact on the environment or human health. For a baseline LCA, where product or process improvement is the major goal, the study conclusions may include a list of areas where the greatest issues exist and greatest improvements can be achieved. Whilst an LCA study does not provide clear-cut solutions to particular problems, through the interpretation of the study findings by experienced professionals the study report may recommend ways in which particular products or processes may be modified to lower these environmental impacts. Any recommendations will be related to the stated goals of the study and the concerns of the intended audience and other stakeholders.

In some comparative studies there may be some instances where one option is not clearly better than the other due to the uncertainties or limitations of the study or the level of importance put on different impacts by different people. Due to the somewhat subjective nature of this phase of an LCA, the conclusions drawn by one person may differ from those of another. In this situation, the results can be used to inform decision makers of the areas where the most significant impacts are occurring. Product or process improvement may become the main use of the study findings and these will be extremely useful in directing this task. The results of this final phase of the study are then presented in a way that meets the requirements as set out in the goal and scope definition phase, depending again on the intended use of the findings and their intended audience.

3.4 How can life cycle assessment be used?

Governments, organizations and even individuals are becoming increasingly aware of the benefits that LCA offers, particularly as society becomes increasingly environmentally conscious and the necessity to reduce our impact on the environment becomes more widely recognized.

A life cycle assessment is capable of providing input into a broad range of everyday environmental decisions. Whilst the initial reasons and intended goals of undertaking an LCA study may vary considerably from one study to the next, the underlying objectives of all studies typically remain the same. These objectives are usually either:

- to help manufacturers, designers, specifiers and consumers identify and/or compare the environmental performance of one or a number of products, processes or services in order to determine the most environmentally friendly option across the entire life cycle of the product or service; or
- to provide a basis for identifying areas of potential improvement in the environmental performance of a particular product or process.

The uses of LCA are broad and, as highlighted by Sonneveld (2000), may include:

- targeting and tracking environmental improvement
- identifying cost and environmental savings from improved resource efficiency
- managing environmental information
- complying with emerging international standards and market requirements
- progressing towards ecologically sustainable patterns of production and consumption
- setting and managing public policy
- providing information for product marketing claims and support for eco-labelling programmes
- providing information for environmentally responsible design.

According to a survey by Cooper and Fava (2006), the main reasons why LCA is used are to support business strategy (18 per cent) and research and development (18 per cent), as an input to product or process design (15 per cent), for education (13 per cent) and for labelling or product declarations (11 per cent).

3.4.1 Environmental improvement

Whilst an LCA can be used for a wide range of purposes, the most useful of these is for product development and improvement. Improvements to the environmental performance of existing company processes or practices may be dictated by the need to comply with external regulations or even internal company policies. This can lead to considerable environmental and financial benefits to an organization. As well as providing competitive advantage over competing products or organizations, providing consumers with environmentally preferred choices also provides a pathway to significant reductions in global environmental impacts.

LCA is often used to compare the environmental loadings of alternative processes within a product system, to assist in identifying the best possible choices of action. Decision makers are then able to adapt existing processes with this knowledge in mind, confident that improvements to environmental performance are actually being made. An example of this may be where a particular technology or process for milling timber can be assessed for the quantity of energy and other resources being consumed as well as the waste being produced per unit of timber output. An alternative milling technology or process that has the potential to improve the efficiency of production (whether from an energy, resources or waste point of view) may then be assessed to determine the potential environmental improvements that are able to be made for the milling process in question. Comparisons of this nature can often involve a detailed level of priority setting and compromise when it comes to selecting the preferred alternative. One alternative may result in lower impacts within one parameter (for example, a reduced quantity of greenhouse gas emissions through a reduction in energy consumption), but a higher level of impact for another parameter (such as an increase in water consumption). This is where the impact assessment and weighting of priorities becomes an extremely important component of an LCA study, as discussed in Section 3.3.3.

However, if an LCA is performed on a particular product or process, as per a baseline LCA, then the objective may be to inform the designer, first, that the product is within a specified limit (and no modification is necessary) or, second, how to modify the product or process in order to decrease its overall environmental impact. The nature of an LCA allows iterative improvements to occur by assessing the performance of these improvements as they are made and modifying the product or processes to further reduce environmental impacts.

3.4.2 Strategic planning

By being able to identify the range of environmental impacts from a particular product or process, decision makers are often in a better position to prioritize improvement strategies and allocate funding in order to maximize any potential environmental benefits. Effort and funding may be targeted towards replacing and modifying inefficient equipment or processes. By linking the improvement strategies to potential benefits, value can then be maximized. This strategy is often aligned with meeting existing and emerging compliance requirements and potential future market demands to ensure that a business stays viable and competitive in an ever-changing market-place with an increasing focus on environmental performance. LCA can be used to identify potential opportunities for a business that may otherwise have been overlooked. These may involve defining strategies for ensuring or growing market share, particularly through improving or emphasizing the environmental attributes of their operations or products; meeting or complying with existing or possible future environmental standards; or minimizing possible future liabilities, risks or costs associated with compliance or shifting market demands.

3.4.3 Public policy making

An LCA can be used to help inform the development, implementation and management of public policy, such as those policies related to environmental performance of the built environment. For example, policy can and has been used to improve the environmental performance of buildings through the establishment of building regulations that deal with their operational efficiency, such as mandatory efficiency targets for heating and cooling (Building Commission 2008). In this case, LCA studies have helped to clarify where significant environmental improvements are needed within the building life cycle, and policy and regulations have been developed to help address these issues.

Ensuring that policy decisions are having the intended outcome and not resulting in unforeseen negative environmental, financial or social implications is also critical. LCA can be used to measure the potential impact that any existing or proposed future policy decisions may have on achieving beneficial environmental outcomes. For example, the implementation of policies that lead to the specification of minimum building insulation levels can be assessed by taking into account and balancing the potential energy and emissions savings against the additional resource requirements associated with production of the insulating materials.

3.4.4 Marketing and eco-labelling

With consumers becoming more aware of environmental issues and demanding products that conform to particular environmental standards or address specific environmental concerns, manufacturers are beginning to use LCA to inform potential consumers of their product's environmental attributes. This can be extremely useful as a marketing tool or to maintain or even enhance their competitive advantage. With *greenwash* becoming more and more prevalent, manufacturers are likely to be increasingly forced into providing reliable evidence to support their claims. LCA can be used to identify potential environmental benefits or attributes of a particular product or process over their own or a competitor's alternative product or process. Many 'green' marketing claims are often based around the findings of such studies and the scientific basis of an LCA is useful for substantiating these claims. An LCA can also be used to demonstrate that particular product improvements have resulted in reduced environmental impacts, such as a reduction in greenhouse gas emissions, and this information may then be presented to consumers through a product declaration or so-called *eco-label*.

In 2009, Wolf Blass Wines introduced their Green Label wine product, which uses a PET bottle in place of the traditional glass packaging. An LCA study was used to compare the greenhouse gas emissions from both forms of packaging, which showed that across the life cycle of the product, the PET bottle produced up to 29 per cent fewer greenhouse gas emissions than the glass alternative (Anon. 2009). Wolf Blass have since been using the findings of the LCA study to promote the plastic product as an environmentally preferred alternative and most importantly to market themselves as a company that is taking an environmentally conscious approach to their operations.

LCA is often used as the basis for environmental labelling and rating schemes, with the findings used to provide environmental information to consumers in simple and easily comparable terms. Examples of environmental labelling include fuel consumption labels that disclose the fuel efficiency of motor vehicles and energy or water rating labels indicating the energy or water efficiency of domestic appliances. The procedures for measuring the consumption figures presented within these labelling schemes are often specified by related standards (for example, International Standard 14025 2006), ensuring that consumers are able to reliably compare competing products on a common basis.

3.5 International LCA standard – ISO 14040 series

The pressure from various environmental organizations to standardize LCA methodology in the early 1990s led to the establishment of a standard on life cycle assessment by the International Organization for Standardization. This standard was first published in 1997, outlining the principles and framework of LCA (*Environmental Management – Life Cycle Assessment – Principles and Framework* (International Standard 14040 1997)). In 1998, International Standard 14041 (*Environmental Management – Life Cycle Assessment – Goal and Scope Definition and Inventory Analysis*) was released, outlining the goal and scope definition and inventory analysis phases. International Standard 14042 (*Environmental Management – Life Cycle Assessment – Life Cycle Impact Assessment*) and 14043 (*Environmental Management – Life Cycle Assessment – Life Cycle Interpretation*), outlining the impact assessment and interpretation phases, respectively, were released in 2000. All of these standards have now been superseded by, and combined into, two standards that were released in 2006: *Environmental Management – Life Cycle Assessment – Principles and Framework* (International Standard 14040 2006) and *Environmental Management – Life Cycle Assessment – Requirements and Guidelines* (International Standard 14044 2006).

These international standards describe the principles and framework behind life cycle assessment and define each of the phases and steps involved in conducting an LCA. Ultimately they provide guidance to those conducting LCA studies and outline several requirements and recommendations to ensure that these studies are conducted in a transparent manner. These guidelines also help to ensure that LCA studies are performed in a consistent manner so that comparative assertions of the environmental performance between different products and processes can be made.

Despite the comprehensive nature of these standards, there remains at least one major limitation to their application. When defining the system boundary and those inputs and outputs to be considered, the standard states that '… resources need not be expended on the quantification of such inputs and outputs that will not significantly change the overall conclusions of the study' (International Standard 14040 2006: 12). As highlighted by Suh *et al.* (2004), it is very difficult to determine whether individual inputs and outputs will have a significant impact on the overall conclusions, particularly prior to data being collected. How can a system boundary be established not knowing whether certain inputs and outputs are going to be significant? The current approach to system boundary selection therefore leaves considerable scope for incomplete system boundaries and hence may lead to the omission of a significant number of the inputs and outputs associated with a product system. One solution to this problem is to use input-output analysis to assist in identifying those inputs that may be most significant. Further data collection efforts can then be targeted towards these areas and greater confidence can be gained from knowing that all significant inputs have been included in the assessed system. This issue is but one of a range of limitations associated with the application of LCA, of which some are outlined in the following section.

3.6 Limitations of life cycle assessment

While other forms of environmental assessment have been in existence for much longer, the most basic forms of LCA were developed in the 1960s in the United States amid concerns over the depletion of raw materials and energy resources (SAIC 2006). Despite this, the awareness of LCA was not widespread in many countries until at least the 1990s. The recent awareness and penetration of LCA in many countries around the world has still not resulted in its widespread implementation (Rebitzer and Schafer 2009). Quite often, LCA is not something that can be easily or rapidly implemented, because of a number of problems and limitations still associated with its use. Many of these problems or limitations result from the subjective nature of the decisions made within an LCA study, such as defining system boundaries, selecting data sources and weighting potential impacts. Some of these limitations are discussed in this section.

3.6.1 Lack of knowledge and awareness

Although concern for the global environment has been growing in recent years, there is still a need to increase the awareness of the impacts that human activity is having on the environment and the actions that can be taken to minimize them. One method of encouraging the use of LCA as a tool for assessing environmental impacts and improving environmental performance, other than financial incentives and legislation, is to educate people, businesses and governments in order to emphasize the environmental consequences of their actions. Decision makers should be made aware that a tool such as LCA could be used to assess these impacts and determine where improvements could be made, or to assist in choosing an environmentally benign alternative. A survey conducted by the Packaging Digest (2008) in 2008 found that of the criteria used to evaluate the environmental sustainability of packaging, only a quarter of respondents were aware of the importance of the life cycle framework that LCA provides. The low levels of awareness found in this study confirm the results of similar previous studies by Berube and Bisson (1992), Miller (1995), Ofori (1998), Tan *et al*. (1999) and Kien and Ofori (2002).

The lack of awareness of LCA in many countries, such as Australia, is mainly due to its relatively recent introduction into these countries. There is no history of its methodological development as in some parts of Europe and the United States and therefore knowledge, experience and understanding is limited amongst academia, government, industry and public stakeholders (Deni Greene Consulting Services 1992; Jensen *et al.* 1997; Evans and Ross 1998; Sonneveld 2000). Due to this lack of awareness, there are difficulties obtaining information and data from various sources, which may either not be available in the required format or not available at all (Sonneveld 2000). However, the awareness and use of LCA is steadily growing as the number of interested stakeholders increases. This is being supported by the realization of the potential for LCA to comprehensively assess the impacts of products on the environment and, in many cases, inform opportunities for providing a competitive edge over competitors.

With a significant increase in awareness of the benefits and uses of LCA amongst industry, government and the general public, further progress may be made towards reducing our impact on the environment (Sonneveld 2000). The creation of awareness and education amongst all stakeholders of the potential benefits of LCA should be the main driver for its further uptake and application across the world (Rebitzer and Schafer 2009). Only in this way will its full potential be exploited.

3.6.2 Methodological gaps

The LCA tool is still very much in a state of development worldwide. Significant conjecture still exists over the most appropriate assessment methodologies, data sources and interpretation techniques that should be used. The most significant issue surrounds the definition of the system boundary. The most common forms of LCA (SETAC and ISO) are forced to draw a boundary around the process(es) being considered, which potentially limits consideration to only a small percentage of the inputs or outputs of the product system, thereby making current assessments incomplete. For example, two LCAs of the same product may be carried out using different system boundaries, and it is possible that they will produce two different results. Treloar *et al.* (2000b: 7) have identified a number of key exclusions when using the most common forms of LCA, including:

- 'ancillary activities associated with the process, such as administration
- inputs of services, such as banking and insurance
- further processing of basic materials into complex products
- non-feedstock energy used to make fuels.'

When undertaking the LCI phase of an LCA, some of the inputs are excluded without consideration or acknowledgment (Berube and Bisson 1992; Lave *et al.* 1995). The system boundary is commonly drawn at a point where it is believed the majority of the inputs to the system have been considered and where any excluded elements could not have any significant impact on the environment. This is a value judgement, usually based on past experience, and is not based on any scientific evidence that any processes further upstream may not be significant. This judgement may not matter when considering a small system or one whose lower-order processes are highly resource-intensive. However, the uncertainties may be much greater when considering a larger system, both in terms of data and associated assumptions (Tillman *et al.* 1994).

The more recent development in the use of input-output data for compiling an LCI allows for the accounting of the upstream processes that have traditionally been excluded and may contribute to a more comprehensive LCI result (Suh and Huppes 2002). Input-output data is believed to be considerably less reliable than traditional process data as it is based on national average data, but it is none the less more complete as it includes a greater range of inputs required for any particular process (Treloar 1997). The use of hybrid LCI methods may help to resolve many of the limitations and issues associated with these more traditional LCI methods.

3.6.3 Geographic issues

Variations in data, potential environmental impacts and the importance of particular impacts can be quite significant across global geographic boundaries. Production and manufacturing processes can vary considerably from country to country, as can the types of fuel used, the source of raw materials, transportation distances and the relevance of specific impact categories. This variation can even be quite significant between areas within the same country. For example, the Australian state of Tasmania relies on low-polluting hydro power for at least 71 per cent of its energy supply (DIER 2009) compared to the other states of Australia that rely heavily on coal and natural gas. Products produced in Tasmania are generally less greenhouse emissions-intensive than their equivalent from other states of Australia, assuming similar technologies and manufacturing processes. Relying on data from other countries or areas that have vastly different production characteristics can result in considerable errors in LCA findings when these are assumed to apply to local conditions. The most accurate and reliable findings are usually obtained where product, process, technology and location-specific data for the particular product being studied is used.

3.6.4 Availability and quality of life cycle inventory data

An LCI involves the collection of a wide range of data from a variety of sources. The level of confidence that can be placed on LCA findings is reliant on the availability and use of appropriate data. Berube and Bisson (1992) and Lewis and Demmers (1996) have identified data availability, its accuracy and the cost of obtaining it as major inhibitors to the use of LCA. The availability of data depends on the extent to which data is being collected and the willingness of other companies to make their data available (Lewis and Demmers 1996). Some data is regarded as commercially sensitive by companies who wish to avoid public knowledge of their impact on the environment. There is a fear that this knowledge may negatively affect their public image or create pressure from authorities to introduce costly measures to reduce these impacts. Often, the release of this data may only occur if the organization is not identifiable or the data is aggregated in a generic form with data from other organizations that produce similar products.

As has been touched upon briefly, many LCA practitioners compensate for the lack of available data by using national averages to obtain a typical description of the product or process being studied. This data is often what is contained within pre-compiled LCI databases providing an aggregation of a number of similar products or processes within a country or region. As such, this data is often not specific to any one product or process. As these processes become more detailed, the range of error between the national average and the product or process being studied becomes greater and greater. Boustead (1996) illustrated the dangers of relying on data that is not specific to the product being studied. For example, there will always be variations in the way in which different manufacturers perform, based on different ages and sizes of plant, the location of the plant, and variations in the way in which the plant is maintained and managed.

The quality of this data is fundamental to the quality and validity of any LCA study (Verbeek and Wibberley 1996; Weidema and Wesnaes 1996). If data used in the LCI phase of the LCA is unreliable, then the results obtained for the downstream phases of the LCA study will only be as good as this data allows. This cumulative error emphasizes the importance of the first two phases of an LCA: that is, defining the correct goals and scope of the study, along with collecting the right data in the LCI, both in terms of quality and depth.

The quality of the data relevant to a specific study will depend on a number of factors, as outlined by Lewis and Demmers (1996), and includes:

- geographic coverage – whether the data is specific to the factory or locality being considered. The impacts of a particular product or process can differ significantly between locations or companies
- time period covered – this must be long enough to average out any significant variations that can occur in data over time
- when the data was collected – the age of the data will determine its accuracy
- technology – the data should relate to the specific technology used in the processes, whether it be old or new. This factor can impact greatly on the LCA outcomes.

As mentioned in Section 3.3.2.2, over recent years LCI methods based on input-output analysis have been developed that attempt to overcome the lack of product-specific data and the problems associated with gathering this data (Lave *et al.* 1995; Pullen and Perkins 1995; Treloar 1997). However, these approaches also have their own problems and limitations, including homogeneity and proportionality assumption and sector-level aggregation.

Homogeneity assumption is the assumption that all of the products produced by a single economic sector are produced in fixed proportions or are direct substitutes for each other, and, moreover, that each sector has a single input structure and there is no substitution between products of different sectors (Carter *et al.* 1981; Miller and Blair 1985; Worth 1993). The sectors into which the economy is divided are homogenous in an economic sense but not necessarily in terms of resource flows (Worth 1993). Although attempts are made to minimize the errors associated with the homogeneity assumption, Pullen (1995) has estimated these errors to be in the order of ±18 per cent.

The use of input-output tables assumes that industries within a sector will pay the same price for products from other sectors. It also assumes that a change in output from a particular sector will lead to a proportional change in the inputs to that sector (Carter *et al.* 1981; Worth 1993). With economies of scale or other changes in the manufacturing processes (e.g. reduced labour), this assumption may become invalid (Worth 1993).

However, most of the problems associated with the validity of input-output data relate mainly to high levels of aggregation (Sebald 1974; Bullard and Sebald 1988; Suh and Huppes 2002). Due to this aggregation, the number of sectors in current input-output tables is very limited in comparison to the vast number of industries, products and services that form part of most economies (Voorspools *et al.* 2000). Furthermore, as input-output tables are not usually designed for assessing the environmental impact of products and processes, a large number of the industries that are associated with significant resource consumption and emissions are highly aggregated (Kohn 1972). Miller and Blair (1985) have estimated the combined errors associated with input-output data to be in the order of ±50 per cent. Despite this, Lenzen (2001: 128) has demonstrated that the errors associated with input-output data are 'often significantly lower than the truncation error of a typical process-based LCA.'

The quality and depth of LCI data, and its gathering, is the basis of any LCA. This collection of data must therefore be carried out with the greatest care and attention if the LCA is to be truly purposeful. In striving to maximize the data quality and minimize problems associated with data collection, a balance must be struck between the use of more reliable process data and the more comprehensive national average input-output data.

3.6.5 Time and cost-intensive

Other problems with LCA, such as the lack of data and low levels of awareness of the LCA tool, have contributed to LCA being time-consuming and expensive to perform. The complex and sometimes difficult task of data collection has a significant impact on the time and costs required to conduct an LCA study (Berube and Bisson 1992). Of the data required for an LCA, very little is publicly available and thus significant effort is needed to obtain the data that is required. Overcoming the data barrier, combined with an increasing knowledge base, will mean that LCA studies should become less time-consuming and therefore less expensive to perform (Sonneveld 2000). As data becomes more accessible, the labour costs involved in conducting an LCA will decrease. Lewis and Demmers (1996) have identified another means of resolving these problems of excessive time and costs. This involves recognizing the 'point of diminishing returns' or the point at which further investment of time or money is not justified by the likely benefits. The point of diminishing returns is where no more time and effort should be allowed than is necessary to inform decision making. The first phase of any LCA – the goal and scope definition – can and should help to define the quality and depth of data that is required, thus determining the necessary time and costs involved.

3.6.6 Interpretation of results

A major concern with LCAs that can have a significant impact on the conclusions and recommendations made is the assessment and interpretation of environmental impacts. Systems that can help to interpret LCIA results are still under development (Lewis and Demmers 1996) and there is no generally accepted methodology for consistently and accurately translating LCI findings into specific potential environmental impacts (Finnveden *et al.* 2009). The interpretation of results occurs at the impact assessment phase of an LCA where inventory items are linked to environmental problems. This interpretation involves value judgements on the relative importance of different impacts, such as water pollution, resource depletion and global warming. A number of decisions need to be made about the priority of these environmental impacts: for example, whether water pollution is more important than the emission of greenhouse gases. Whatever decisions are made, these, along with any assumptions, should be clearly acknowledged and explained (Lewis and Demmers 1996).

Results from an LCA provide an input into a wide range of decision-making processes and the findings can be, and are, used for a wide range of purposes. However they do not necessarily provide a definitive answer to specific questions. The findings are only indicative of the potential impacts that may occur across the life cycle of a product or process (International Standard 14040 2006: 9). The relevance and usefulness of any findings are determined by not only how well the study has addressed the initial goals, but also the experience and ability of the LCA practitioner to interpret these findings.

Despite these current issues and limitations, the LCA field has undergone rapid advancements over the past two decades and continues to develop and progress. This work will help to ensure that in the future many of these current limitations are able to be minimized or even completely avoided.

3.7 Summary

Life cycle assessment is but one of a number of tools or techniques that can be used to assess the environmental performance of a range of products or processes. The strengths of LCA are numerous and include the ability to identify opportunities for improvement in environmental performance, compare competing alternatives and inform decision making associated with the design process all the way through to government policy setting.

The assessment process can be as simple or as complex as is necessary to address the intended use of a study's findings; however, the practitioner must be aware of the potential issues and limitations that this may cause when relying on these findings for key assertions. Of the four key phases of an LCA, the data collection process is critical to ensuring that a reasonable level of confidence can be given to the study findings and any claims or comparisons that are made based on these. The quality, reliability and usefulness of study findings can easily be compromised by poor data quality, a lack of data, choice of assessment approach or incorrect or unrealistic assumptions.

Despite the relatively long history of the development and application of LCA and a set of International Standards describing the LCA framework and its application, there remain significant impediments to its widespread use. Much of this stems from a continuing debate over appropriate assessment methodologies, availability of data, complexities of assessment and time and cost constraints.

This chapter has provided an overview of the main steps involved in conducting an LCA, as well as the most common uses for LCA studies. LCA is able to provide detailed input into a broad range of environmental decisions, informing product improvement, consumer choices, policy making, eco-labelling and marketing and business strategies.

The use of LCA for assessing the various elements of the built environment is crucial for improving the environmental performance of these elements due to their complex nature and the amount of time that much of the built environment remains in service. Using LCA in the design and management of the built environment can lead to significant improvements in its environmental performance. The application of LCA in this context can also be extremely time-consuming and involved, and a considerable commitment is required from all involved stakeholders to ensure the potential benefits of using LCA are realized.

4 Quantifying environmental impacts of the built environment

This chapter describes the application of life cycle assessment (LCA) for quantifying the environmental impacts associated with the built environment. The assessment process is described with specific reference to buildings. Particular emphasis is placed on streamlined LCA and input-output-based hybrid life cycle inventory analysis, as these approaches are considered to be most relevant to identifying and assessing the environmental impacts associated with the built environment. The use of LCA within the built environment is most beneficial when used during the initial design phase. However, due to the considerably long life cycle of buildings and many of the other elements within the built environment, changing environmental priorities and issues and technological advances over time mean that the use of LCA at the other key life cycle stages is also crucial and can be extremely useful in decision making.

A residential building case study is used throughout this chapter to demonstrate the application of the various stages of an LCA within the built environment. This example demonstrates the various steps involved in collecting and analyzing data, converting this LCI data to environmental impacts and interpreting the findings of an impact assessment. Of particular emphasis is the application of the innovative input-output-based hybrid approach for conducting an LCI.

There are numerous software tools available to help facilitate the LCA process within the built environment. Some of the tools are specific to the assessment of buildings, whilst others can be used for assessing the environmental impacts of a much broader range of products. These tools also consider a variety of environmental parameters and impact categories. A number of these tools, including the extent to which they cover the broad range of life cycle stages and environmental parameters, are described briefly at the end of this chapter.

4.1 Life cycle thinking in the built environment

Buildings and other elements of the built environment are often built to last many decades. They also involve consumption of resources and impacts across a variety of stages during their life, as described in Chapter 1. For example, the

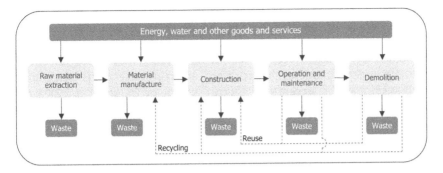

Figure 4.1 Inputs and outputs across the building life cycle

construction of buildings consumes enormous quantities of raw materials, energy and water and results in the release of a variety of pollutants, greenhouse gases and waste, both during the construction stage and associated processes (such as material manufacture, transportation etc.). The operation of buildings once occupied also consumes large quantities of energy and water and produces considerable quantities of waste over the building's life. The eventual demolition or reuse of these buildings is also not without requirements for energy, water and other resources. The range of inputs and outputs across the various stages of the building life cycle are depicted in Figure 4.1.

Buildings are unique in that their size, complexity and resource-intensive nature across all stages of their life cycle mean that their impact on the environment stems from a large range of disparate processes, systems and requirements and is also heavily influenced by factors such as human behaviour. The relatively long expected life of buildings and many of the other elements within the built environment poses a problem when it comes to assessing the potential environmental impacts that may result from each of these life cycle stages, particularly as these may occur many decades from when the building is initially constructed. However, for elements such as buildings where impacts are often spread across many stages and years, it is crucial that a life cycle approach is used in order to assess and ultimately minimize these impacts.

4.1.1 Why is a life cycle approach important?

Many environmental performance studies define the scope of analysis to include only a limited number of impacts, processes or stages. This streamlined approach (as discussed in more detail in Section 3.2.3 and 4.2.2) may be well warranted where evidence exists that provides an indication of significant impacts occurring from a particular stage, process or impact category. Often though, considerable effort is placed on these elements on the basis of incorrect assumptions that these impact categories, processes or life cycle stages are where the major issues lie. For buildings, this can be evidenced by the considerable effort placed on assessing and reducing impacts from building operation and the lack of consideration of, and effort directed towards, construction-related impacts. Some of these deficiencies may be explained by a lack of comprehensive assessment methodologies and a previous underestimation of impacts from certain processes or life cycle stages, such as construction. Recent studies (such as those by Treloar *et al.* 2000a and Crawford 2008), using the comprehensive LCI assessment technique known as input-output-based hybrid analysis, suggest that the energy embodied in constructing buildings can be equivalent to the energy required for their operation over their life. Previous studies, such as those by Hill (1978) and Cole and Kernan (1996), using incomplete LCI assessment techniques indicated that the embodied energy component of a building may represent less than 10 per cent of its lifetime energy consumption. The assumptions made, and the methods and approaches used, can therefore have a significant impact on the overall findings of any LCA study. It is important that the LCA practitioner is aware of the full life cycle environmental implications of any product or at least has an awareness of the potential limitations of the assumptions made and assessment methods chosen.

Often, changes or improvements to particular processes or life cycle stages can unknowingly have a flow-on effect on other processes or life cycle stages, both positively and negatively. For example, improvements to the operational efficiency of buildings through the addition of thermal insulation or thermal mass (in the form of bricks or concrete, for example) also results in an additional requirement for materials and hence additional energy, water and raw materials for their manufacture. An example of this is in Council House 2 (CH_2) – the City of Melbourne's new office building – where numerous strategies have been employed to reduce operational cooling energy requirements to around 83 per cent less than that of a conventional building of this type and size, which is equivalent to 280 GJ per year (City of Melbourne 2006). One strategy that contributes to this saving is the use of a phase change material encased within 23 tonnes of stainless steel balls that helps to cool down the water being used within the building's cooling system. The inclusion of these balls has added to the quantity of material that would otherwise have been required for a conventional cooling system and thus the impacts of this must be balanced against the benefits that it provides. The energy required to manufacture the balls alone equates to approximately 10,190 GJ (based on an embodied energy coefficient of 443 GJ/t for stainless steel (Treloar and Crawford 2010)). Conservatively, assuming that this system itself is contributing to half of the cooling energy savings of the building, this would equate to an energy payback period (or the time it would take for the energy embodied in the stainless steel balls to be offset by the operational energy savings) of over 70 years, potentially well beyond the life of the system itself and possibly even the building.

Taking a life cycle approach, where all life cycle stages are considered and the potential effects of changes in each stage on each and every other stage are assessed, is crucial to ensuring that the environmental performance of a

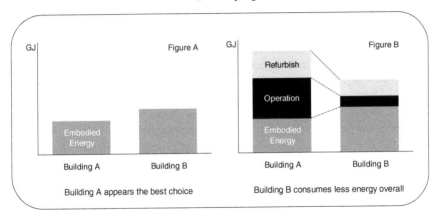

Figure 4.2 The importance of a life cycle approach in the built environment

building or any other product is optimized across its life cycle. This approach ensures that the multiple environmental and resource issues across the entire life cycle of the building or any element of the built environment are identified. In this way, any reduction in impacts at one point does not simply create greater impacts at another point in the life cycle.

Figure 4.2 illustrates the importance of taking a life cycle approach to environmental improvement. Figure A depicts a situation where only the energy-related construction impacts have been considered, showing Building A as the better option from the perspective of reducing embodied energy associated with construction. Figure B shows what might happen when the energy requirements associated with the other stages of a building's life are considered (namely operation and refurbishment). The higher initial embodied energy of Building B is a result of the specification of more durable, longer-lasting materials and components, which reduce the frequency and resource requirements of maintenance and replacement, as well as higher levels of insulation and more efficient systems, which reduce the energy associated with building operation. The results show that the net energy benefits of these improvements mean that Building B is actually the preferred option for reducing life cycle energy requirements. Life cycle improvement is a much more favourable outcome but can only be assured if it forms part of the early-stage decision-making processes.

4.2 Using life cycle assessment in the built environment

The construction, operation or use and eventual demolition of buildings and other elements within the built environment contribute towards a broad range of impacts on the natural environment, ranging from global warming, waste production, depletion of natural resources, pollution of air, land and water and also human health impacts. Suzuki and Oka (1998) believe that given the varied materials used in the construction, operation and demolition of buildings and other elements of the built environment, the potentially enormous range of environmental criteria that are relevant to these items may serve as a severe limitation to the use of LCA methods for assessing their environmental performance. Buildings are more complex than many other products in that they involve the aggregate effects of a host of life cycles of their constituent materials, components, assemblies and systems (Cole 1998). However, as noted in Chapter 1, some of the most significant impacts from the built environment can be attributed to the consumption of fossil fuels. As buildings are particularly energy intensive across their life, existing evidence suggests (for example, Junnila and Horvath 2003) that this fossil fuel consumption is the most important issue in addressing their impact on the environment and the reason why most LCA studies in the built environment have energy use as their main point of focus. For elements such as buildings, where existing evidence suggests that the most significant environmental issues are centred on a limited range of impacts (such as those attributable to fossil fuel consumption), a limited scope for an LCA study may be warranted. This can be achieved through a streamlined LCA approach.

4.2.1 Supporting decisions across the life cycle

LCA can be useful at the design stage for assessing the potential environmental impacts that may result from particular design decisions or choices. It can also be a useful tool for assessing environmental impacts associated with operational improvement opportunities, refurbishment or maintenance strategies or demolition and disposal strategies for elements of the built environment at any stage during their life. This is particularly useful for buildings, for example, as new construction each year adds less than 2 per cent to the existing building stock (Holness 2008). A key opportunity for achieving rapid efficiency gains and environmental improvement in the built environment exists in the many existing buildings that are often underperforming from an environmental perspective. Due to the long and complex life associated with these buildings, many of the decisions associated with their original design and construction are still having considerable environmental consequences years or decades after they were built.

Whilst the use of LCA during the design phase is important and can help direct and inform design decisions, there are no guarantees of what the environmental impacts will be, especially for those resulting from the life cycle stages furthest into the future (such as refurbishment and disposal). This is why it is often important that an LCA is conducted at each of these key life cycle stages. Assessing buildings and other elements of the built environment at key stages of their life cycle can be extremely beneficial. It ensures that the intended environmental outcomes defined during the initial design stage have been and are being achieved, assists in decision making needed at each of these stages and identifies opportunities for further environmental improvement in light of changing priorities, demands and technologies over time. For example, many existing buildings perform poorly from an energy and water efficiency perspective as they were designed well before the costs and environmental advantages of improved operational efficiency were considered of universal importance. By assessing the environmental performance of these existing buildings, opportunities for improved operational performance can be identified. The most obvious of these opportunities is often a reduction in energy consumption through the upgrade of existing lighting or HVAC systems. LCA can then be used to assess various options for addressing these issues, as well as other refurbishment strategies, to ensure that net environmental benefits are being achieved.

The value of resources changes over time and many of the raw materials used to produce existing construction materials may either be depleted or close to depletion by the time our current new buildings come to the end of their physical life. As such, those environmental issues seen as priorities now may be overshadowed by more pressing environmental concerns in the future. For example, the recovery of certain materials at the end of a building's life may not currently be considered a high priority due to an abundance of raw materials or other environmental issues associated with the recovery process. As these materials become scarcer and their value greater, their imminent depletion may become a more important issue.

4.2.2 A streamlined life cycle assessment approach

A streamlined approach to LCA involves limiting the scope of the study in any number of ways. This may involve an analysis of a limited range of life cycle stages (such as the construction or operational impacts), environmental parameters (such as energy or water), impact categories (such as global warming or resource depletion) or processes (such as transportation or manufacturing of specific materials or components). The decision to take a streamlined approach may be informed by a number of different factors: the reasons for undertaking the study, which may be informed by regulatory, marketing or other internal or external demands; a previous full LCA may have identified particular areas where significant issues exist, and future improvement efforts can then be targeted; or the goals or intended application of the findings relate to a particular environmental parameter, life cycle stage or impact category or the importance of particular environmental issues. Whilst limiting the scope in this way may be well warranted and acceptable, it is important that the fundamental advantage of LCA and its ability to assist in avoiding sub-optimization is not compromised – for example, by addressing one issue that then leads to another problem elsewhere in the life cycle – as this may lead to serious flaws in the results obtained. A streamlined approach can also be useful to provide an initial indication of where some of the most significant issues lie. For example, does the energy consumed during the operation of a building represent a greater or lesser issue than that consumed during the construction phase? Similarly, a streamlined approach may be used to determine whether water or energy consumption is of greater importance during the operational stage of a building's life cycle.

4.3 Goal and scope definition

As for any LCA, it is important that the goals and scope of a built environment-focused LCA study are defined in relation to the intended use of the results. This will help to ensure the applicability and usefulness of the study findings and will prevent unnecessary time and costs being spent on data collection and analysis that are outside of the project's intended purpose.

4.3.1 Goals

The goals of an LCA in the built environment can be guided by the need for environmental improvement of individual products or processes, to ensure compliance with standards or regulatory requirements or to provide information for product marketing claims. They also need to relate to the specific application of the findings and the intended audience.

Due to the complex nature of many of the elements within the built environment, specific goals can be many and varied, from providing information for building material or product manufacturers to substantiate marketing claims, particularly on the basis of environmental performance; to improving the performance of internal manufacturing processes to reduce environmental impacts and comply with internal environmental policies; to providing information to building designers for specifying building components; or to providing information to inform and direct the development of government policy.

The remainder of this chapter uses a residential building case study to demonstrate some of the steps involved in applying LCA within the built environment. In order to simplify this, a streamlined approach has been used, as defined in the following goal and scope definition phase. The details of the case study are shown in Figure 4.3.

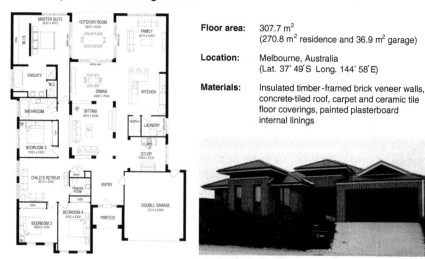

Floor area: 307.7 m^2
(270.8 m^2 residence and 36.9 m^2 garage)

Location: Melbourne, Australia
(Lat. 37° 49′ S Long. 144° 58′ E)

Materials: Insulated timber-framed brick veneer walls, concrete-tiled roof, carpet and ceramic tile floor coverings, painted plasterboard internal linings

Source: Metricon 2010 (plan).

Figure 4.3 Floor plan, front elevation and details of residential building case study

---

RESIDENTIAL BUILDING CASE STUDY – GOAL DEFINITION

The purpose of the study is to identify the global warming impacts from the construction and operation stages of a residential house located in Melbourne, Australia, in order to identify and prioritize strategies for environmental improvement across these two stages.

4.3.2 Scope

A full building LCA will include an assessment of a broad range of impacts resulting from all resource requirements or inputs (raw materials, energy, water etc.) and all outputs (waste, pollutants and emissions) from every stage of a building's life (raw material extraction all the way through to final disposal). As noted previously, such an assessment for complex products such as buildings would be extremely time, resource and cost-intensive. A streamlined approach, focusing on those issues or life cycle stages known to be significant or of most importance, is often ideal in this situation. The scope of such a study can be defined in a number of different ways, limited by either the range of environmental parameters, impact categories or life cycle stages considered, as Table 4.1 shows.

Table 4.1 Possible scope limiting factors in a life cycle assessment

Scope limited to:	Example
Environmental parameter	Energy or water consumption
Impact category	Global warming or resource depletion
Life cycle stage	Construction phase

More often than not, an LCA is limited in scope in one way or another, either intentionally or without knowledge. It is when the scope has been limited without the practitioner's knowledge, such as by decisions that have been made about the source or type of data or assessment approach to be used, that it is of greatest concern. In the worst case, this may lead to decisions that are based on incomplete or unreliable data and potentially incorrect findings.

4.3.2.1 Functional unit

The built environment consists of physical products or items that each have their own individual purpose. The functional unit chosen can vary across these different products dependent on how the product is used and the basis on which relevant comparisons are best made. For example, buildings are often best assessed on a per square metre basis which makes it easier for comparison to other buildings. Items of infrastructure are often best assessed based on their performance characteristics, such as for roads on a per kilometre basis over a specified number of years.

4.3.2.2 System boundary

The system boundary is defined by either the goals of the study, the ability to obtain the necessary data or the perceived significance of individual processes, impacts or life cycle stages. The boundary may cover just one process (such as transportation of materials from factory to site), one life cycle stage (such as the operation of buildings) or one impact (such as global warming), or any number of each of these and in any combination. For example, a study that aims to provide information that will help to reduce the energy-related impacts from building construction may limit the assessed system boundary to energy use from only those life cycle stages upstream of and including the main construction process.

RESIDENTIAL BUILDING CASE STUDY – SCOPE DEFINITION

The study will quantify the energy inputs associated with the construction (including all supporting inputs) and operation of the house over a 50-year period (in gigajoules). Energy will be converted to primary energy terms where direct energy consumption figures are collected to account for the impacts associated with its production. Energy figures will then be converted to carbon dioxide equivalents (CO_2-e).

Functional unit: The product to be analyzed is a 307.7 m² single-storey detached house (as detailed in Figure 4.3) located in Melbourne, Australia.

System boundary: Only the construction and operation stages will be assessed, excluding energy required for maintenance, refurbishment, demolition, reuse and recycling.[2] All energy requirements associated with the actual construction process and all supporting processes and services upstream of this will be included, such as, but not limited to, transportation and manufacturing of materials and the provision of capital equipment (the initial embodied energy of the house). All of the energy required for operating the house (for heating, cooling, lighting, appliances, hot water, cooking, refrigeration etc.) will also be included. The energy embodied in furniture and non-fixed appliances (such as washing machines and refrigerators) will not be assessed. Figure 4.4 indicates the system boundary chosen for the study.

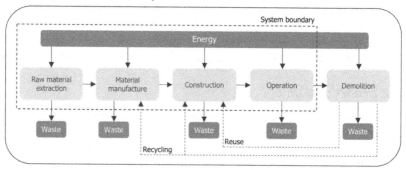

Figure 4.4 System boundary selection for residential building case study

Impact category: Global warming
Category indicator: Global warming potential (CO_2-e)

4.4 Life cycle inventory analysis

Based on the goals and scope of the study, the inventory analysis phase will involve collecting data necessary to inform the study's intended purpose. This data is then validated and converted to the appropriate unit or format to match the functional unit being assessed.

4.4.1 Quantifying inputs and outputs

Inputs may include consumption of energy, water, raw materials (e.g. iron ore for steel production or limestone for cement production) or products and will vary depending on the product or processes being assessed. The outputs that may be quantified include volumes of emissions to the air and discharges to soil and water, as well as the quantity of waste produced. Data should be collected in the most appropriate units for the input or output being measured. The units of each type of data will vary from units of energy or water consumed (e.g. gigajoules (one GJ = 10^9 joules) and kilolitres (kL)) to a volume or mass of raw materials, waste or emissions (e.g. cubic metres or tonnes).

Data will usually be collected from production sites or other sources. For buildings and other items of infrastructure common to the built environment, this will predominately be from materials and product manufacturers such as metal producers, concrete plants and glass manufacturers. Collecting data from specific manufacturers or processes should be the highest priority as this data typically represents the most reliable data available for specific products. This data, commonly referred to as *process data*, relates directly to an analysis of individual and specific processes. Collecting this data, however, is a very time-consuming task as much of the data will sit across extended supply chains. Difficulties can also arise when trying to obtain environmental data from individual organizations as this information is often considered to be commercially sensitive.

Data should be collected for each process that is part of the system boundary and may be either measured, calculated or estimated depending on availability. Environmental Performance Data Sheets (EPDS) are often used to facilitate the data collection process and ensure that data is collected in a consistent manner across potentially hundreds of processes (see Centre for Design (2001) for an example of an EPDS). This data then forms the basis of a product's environmental declaration or environmental profile (International Standard 14025 2006). Whilst a consistent format for presenting this data is useful, especially for avoiding double-counting issues and for comparing similar processes between different manufacturers for example, the most time-consuming and difficult task is sourcing the data required. Whilst data on some inputs and outputs may be routinely collected or available as part of normal business practices (energy and water consumption or greenhouse gas emissions, for example), identifying each and every supplier upstream in the supply chain and collecting the data associated with their processes, and then the processes used by their suppliers and so on, can be extremely difficult, if not impossible.

The reliability and thus usefulness of the LCA results will depend on the quality of the data collected. The two most important indicators of data quality are its reliability and completeness. The reliability of the data indicates how well the data reflects the actual processes being analyzed. A measure of the data completeness indicates how well the data covers all of the processes within the system boundary. If significant gaps exist then further data may need to be collected. Using a complete system boundary avoids the possibility of leaving out any potentially significant inputs. As highlighted in Chapter 3, traditional LCI methods are often fraught with problems that restrict the ability to achieve either reliability or completeness.

RESIDENTIAL BUILDING CASE STUDY – DATA REQUIREMENTS

Embodied energy: Energy requirements for constructing the house will be based on the energy required to manufacture the individual construction materials. A bill of quantities for the house will be compiled (based on the architectural plans) itemizing each material used in its construction. Process data will be sourced from the SimaPro Australian database for materials, giving quantities of energy required per unit of common building materials. Input-output data will be used to fill the upstream data gaps for these materials. Any further data gaps will be identified (for those inputs not covered by the process data) and filled using input-output data in accordance with an input-output-based hybrid LCI approach (see Section 4.4.2).

Operational energy: The energy required during the operation of the house will be determined from existing energy bills (gas and electricity). Bills for an average year will be used and extrapolated to determine the total operational energy requirements over 50 years.

4.4.2 Compiling a life cycle inventory using input-output-based hybrid analysis

Based upon the goals and scope of the study and the data types known to be needed and available, as well as the potential source of this data, information on the inputs and outputs of the product system can be quantified.

As discussed in Chapter 3, an input-output-based hybrid approach provides the most comprehensive system boundary possible for calculating the inputs and outputs for a product system, avoiding or limiting many of the problems associated with other LCI methods, such as system boundary incompleteness. This approach is well suited to assessing the environmental impacts occurring within and from the built environment as it can easily deal with its complex nature and the fact that many of the most significant impacts occur a number of stages upstream of the main construction process.

4.4.2.1 Input-output analysis

A disaggregated input-output model (showing the magnitude of inputs and outputs for each individual process associated with any product or service within any economic sector) can be an extremely useful tool for providing an initial indication of the significance of individual flows within the supply chain of a particular product or process (for example, building construction). For example, these models can be used to identify which processes demand the greatest quantity of inputs (such as energy or water) or produce the greatest quantity of outputs (such as waste or emissions). An energy-based input-output model developed by Prof. Manfred Lenzen at the University of Sydney (Lenzen and Lundie 2002) shows that the greatest quantity of energy consumed for a single process associated with the construction of a residential building is for the manufacture of ceramic products (including clay bricks, ceramic tiles and other ceramic items within a house). Figure 4.5 shows the ten highest energy-consuming processes (or pathways) for the Australian *Residential building* sector, representing 23 per cent of the *total energy requirement* of the sector (a list of the top 100 energy pathways, representing 37 per cent of the *total energy requirement* of the Australian *Residential building* sector, is given in Table A.1, Appendix A). Each one of these processes represents a specific product or service required during (or embodied in) the building life cycle stages upstream from and including the main construction process. The production of clay bricks alone requires energy to fuel the machinery used to extract the raw materials from the ground (such as clay), to fuel the trucks that transport these raw materials to the manufacturer, to power the kilns and other equipment used to produce the bricks and to fuel the trucks to transport the finished product to site.

In Figure 4.5, second-order processes, represented as *Sector A* into *Sector B*, indicate a Stage Two energy requirement (two stages upstream of the construction process). These include, for example, the energy required in producing steel that is used by the *Household appliances* sector to produce a washing machine (*Iron & steel* into *Household appliances*).

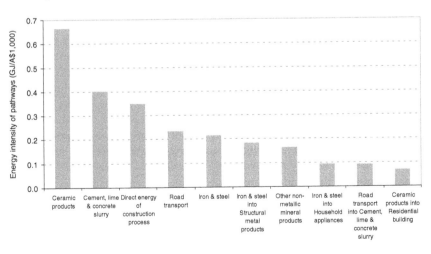

Figure 4.5 Top 10 energy pathways from the Australian *Residential building* sector, based on 1996-97 input-output tables

This initial step can be extremely beneficial in targeting further (process) data collection efforts for processes that are considered to be significant and in avoiding excessive time and money being spent on collecting data for processes that may be insignificant. The ranking of important pathways for the *Residential building* sector in Figure 4.5 indicates that efforts would be best spent on ensuring that data for ceramic products, concrete and steel production covers these processes in as reliable and comprehensive a manner as possible. This approach also provides valuable information that may be used to refine the scope of the LCA study, informing particularly the processes to be covered and the types and quality of the data required to cover these processes.

Input-output analysis can also be used to provide an initial indication of the magnitude of the total requirement for energy or other inputs for any product or service within an economy based on the *total energy requirement* of the sector (the total quantity of energy required to produce an average product from the sector in units of input (e.g. gigajoules of energy) per dollar of sector output). For the Australian *Residential building* sector, this equates to 10.633 GJ per $1,000 of building cost, meaning that it takes 10.633 GJ of energy to produce $1,000 worth of building construction.

4.4.2.2 Process data collection

Much of the process data relevant to the built environment that is typically collected relates to the manufacture and transportation of major building materials (such as concrete, steel, plastics and glass). The system boundary covered is generally limited to the main processes (e.g. manufacturing and transport) and a limited number of inputs upstream of this (such as fuel required for the machinery used to extract raw materials, raw material processing and intermediate transport requirements).

Whilst much of this data is collected on a project by project basis and not usually available to the general LCA community, some process data can be found in the numerous commercial software tools and compiled databases available around the world. This data is often aggregated to varying levels of detail, sometimes combining the data from a number of different manufacturers of a similar product within a country. The applicability of such data for assessing the inputs and outputs of a product produced by a specific manufacturer is thus limited, although significant time and cost savings can be made where the national average data is sufficiently representative of the processes being assessed and it may not be necessary to collect this data again. One solution to this may be to replace only that data for particular processes or inputs where the product is considered to perform better or worse than that of the average covered by the common process data. This type of data can be particularly useful where there are only minor variations in manufacturing processes and technologies across the producers or suppliers of particular products within a country. Caution should also be exercised when using data from a particular manufacturer as this data is typically relevant to only specific manufacturing processes, technologies or situations. The use of such data limits the applicability of the findings outside of these circumstances as significant variations can occur between different manufacturers.

4.4.2.3 Integrating process and input-output data

The individual processes included in a *process analysis* typically include only a small proportion of the total number of those associated with any product system. Whilst individually the majority of the remaining processes represent only a small proportion of the total inputs or outputs of a product system, their sum can represent up to 87 per cent of the inputs or outputs associated with any product system (Crawford 2008). This means that a *process analysis* alone may exclude up to 87 per cent of the energy or other resource requirements or emissions associated with a product. Depending on the profile of the supply chain this percentage can vary significantly. For products that involve very resource-intensive lower order processes (such as the production of aluminium, steel, cement and glass), it may be possible to cover a much greater proportion of their total resource requirements, as process data for direct (Stage Zero) or Stage One inputs is generally more readily available than data further upstream in the supply chain. In some countries where process data is available in much greater quantities and detail, it could theoretically cover more than half of the inputs or outputs to the product system.

The complexity of the supply chain becomes more apparent the further upstream of the main process one goes. It is often very difficult and time-consuming, if not impossible, to collect process data for most of the upstream processes beyond Stage One (one transaction upstream of the main process). However, the potential significance of the inputs and outputs for these processes, particularly for the three sectors covering buildings in the Australian input-output tables (*Residential building*, *Other construction* and *Prefabricated buildings*), is evidenced in Figure 4.6, showing the proportion of each sector's *total energy requirement* at each stage upstream of building construction. The sum of energy inputs upstream from and including Stage Three represents at least 44 per cent of the *total energy requirement* of each of these three sectors.

Even many of the processes that directly support the main construction process (those at Stage One) are often omitted from a process analysis. There are a number of reasons why these processes are typically not covered, including a lack of awareness of their existence and the difficulties (practical, financial and time-related) associated with collecting data for these processes. A number of the Stage One processes that have the most significant energy inputs for the Australian *Residential building* sector, typically excluded in even the most comprehensive process analyses, are shown in Table 4.2. These processes cover those goods and services purchased directly by the construction company in most cases.

To ensure that the system boundary covers as many of the relevant inputs and outputs as possible and any truncation at any point in the supply chain is avoided, it is important that the input-output model forms the basis of the data collection efforts. Process data can then be integrated into this model where it is available and the input-output data can be used as the best representation of the remaining inputs and outputs for the product system.

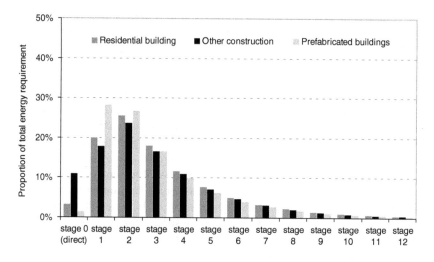

Figure 4.6 Total energy requirement at each stage as a proportion of total energy requirement for the Australian construction-related industry sectors

Table 4.2 Most significant energy inputs of the Australian *Residential building* sector not typically included in a process-based life cycle assessment

Input sector	TER (GJ per A$1000)	Proportion of sector TER (%)
Residential building (repair and maintenance)	1.1090	10.43
Other property services	0.3548	3.34
Wholesale trade	0.3509	3.30
Direct (construction process)	0.3490	3.28
Other machinery and equipment	0.0832	0.78
Agricultural, mining and construction machinery	0.0725	0.68
Accommodation, cafes and restaurants	0.0647	0.61
Legal, accounting and marketing services	0.0611	0.57
Other business services	0.0481	0.45
Mechanical repairs	0.0471	0.44
Total for top 10 inputs	2.5404	23.88

TER = Total energy requirement.

REPLACING INDIVIDUAL PATHWAYS WITH PROCESS DATA

The key to integrating process data into an input-output model is to be able to identify which of the processes from the input-output model the collected process data represents. Process data will typically represent inputs or outputs for a process at the first or second stage upstream of the main process (such as construction), but may, depending on the availability and detail of data, include inputs further upstream in the supply chain.

Let us assume that the following process data was available for the construction of a concrete slab used in a residential building:

- Energy use for the concrete production process
- Energy use for the cement production process (excluding clinker and gypsum processing)
- Energy use for the production of gravel.

An understanding of the product being assessed and the economic sector into which the product is classified is crucial to identifying the input-output model pathways to be replaced with process data. This level of detail is usually provided with published input-output tables as the Input-Output Product Classification (IOPC), which details the individual commodities (goods and services) and the sector into which they are categorized (ABS 2001b).

A disaggregated energy-based input-output model used to identify the flow of discrete energy pathways between all economic sectors (and thus the aggregated goods and services purchased and sold to produce or supply any other goods or services) provides an indication of the magnitude of all of the upstream energy inputs required to provide concrete for residential building construction. This shows that for a threshold of 0.0001, there exists 545 individual energy inputs associated with the production of the concrete components of a residential building (such as a floor slab). The top 10 of these, by the *direct energy requirement* of the individual processes, are shown in Table 4.3. For example, item two (*Road transport*) represents the energy required for transportation of concrete from production plant to building site, whilst item seven (*Road transport → Other mining*) includes the energy required for transportation of gravel from the quarry to the concrete production plant.

From this breakdown, the pathways represented by collected process data can be deduced. The cement and gravel production processes occur one stage upstream of the concrete production process (as they are direct raw materials required for concrete production), so these pathways are represented as in Figure 4.7.

Cement and (ready-mix) concrete are both included within the *Cement, lime and concrete slurry* sector of the economy, whilst gravel is classified as part of the *Other mining* sector.[3] Therefore, the sectors representing the production of each of these materials required by the *Residential building* sector to produce a concrete slab are as follows (as per Table 4.3): for concrete,

Table 4.3 Top 10 energy inputs to concrete production for the Australian *Residential building* sector

Input sector	DER (GJ per A$1000)
1 Concrete production process	0.4018
2 Road transport	0.0922
3 Cement, lime & concrete slurry	0.0436
4 Other mining	0.0194
5 Road transport → Road transport	0.0103
6 Road transport → Cement, lime & concrete slurry	0.0100
7 Road transport → Other mining	0.0049
8 Cement, lime & concrete slurry → Cement, lime & concrete slurry	0.0047
9 Rail, pipeline & other transport	0.0046
10 Other non-metallic mineral products	0.0027
Total for top 10 inputs (60% of TER of concrete production)	0.5942

DER = direct energy requirement.

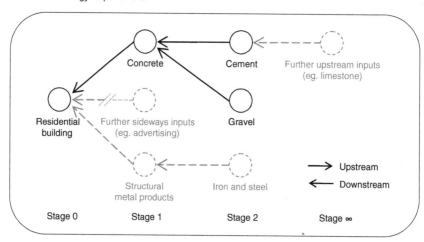

Figure 4.7 Residential building system boundary indicating inputs to concrete production

the concrete production process itself (input one); for cement, *Cement, lime and concrete slurry* (input three); and for gravel, *Other mining* (input four).

As the energy requirements from the energy-based input-output model are generally given as a quantity of energy input per dollar of product output, these values must be converted to an energy requirement associated with the concrete slab, based on its cost. This is simply done by multiplying these values by the price of the product. So, for a residential building with a construction cost of A$160,000, Equation 4.1 would be used to determine the *total energy requirement* (TER) associated with its construction:

$$TER = \$160,000 \times 10.633 \text{ GJ per A\$1,000 }^{4} \qquad\qquad \text{Equation 4.1}$$
$$= 1,701 \text{ GJ}$$

The energy requirements of the individual items that support the building's construction can be calculated in the same manner. Table 4.4 shows the input-output-based energy requirement for the concrete, cement and gravel production processes for a A$160,000 residential building, based on the input-output model.

Process-based energy requirement values for a particular building are determined based on the quantity of specific materials contained in that building and the quantity of energy required to manufacture a unit of those materials (e.g. a cubic metre of concrete), which is also known as the material's energy coefficient. Assuming that the only concrete used for the A$160,000 residential building is in the floor slab and that the floor area of this building is 250 m², this would equate to approximately 70 m³ of concrete. Process-based energy coefficients for these three processes associated with concrete production are given in Table 4.5 in units of gigajoules (GJ) of energy per cubic metre of concrete (based on Grant 2002). These coefficients are multiplied by the quantity of concrete required (in this case, 70 m³) to determine the total process-based energy requirement for each of these three processes (as shown in Table 4.4).

Table 4.4 Process and input-output direct energy requirement (DER) values of selected concrete production processes for A$160,000 residential building

Process	Process DER (GJ)*	I-O pathway DER (GJ/A$1000)	I-O DER (GJ)
Concrete production	25.2	0.4018	64.28
Cement production	44.1	0.0436	6.98
Gravel production	3.5	0.0194	3.10
Total	72.8		74.36

*Calculated based on process-based energy coefficients from Table 4.5.

Table 4.5 Process-based energy coefficients for the direct energy requirement of selected concrete production processes

Process	Process DER (GJ/m³ concrete)
Concrete production	0.36
Cement production	0.63
Gravel production	0.05

Source: Based on Grant 2002.

The input-output value (I-O DER) for each of these processes is then deducted from the *total energy requirement* of the *Residential building* sector (10.633 GJ per A$1,000) and replaced with the process data values (Process DER) for these processes. The result of this process is a complete system boundary covering all inputs of energy associated with the construction of the building, with more reliable process data used to replace the equivalent input-output data where it is available (in this case only for the concrete, cement and gravel production processes).

Further pathways in the input-output model may also be replaced with process data where it is available. For example, if the energy required to manufacture the steel used for the reinforcement of the concrete slab was also available, then the *direct energy requirement* for the *Iron and steel →Structural metal products* pathway (at Stage Two in Figure 4.7) would be replaced with the available process data. However, the *direct energy requirement* of the *Structural metal products* pathway itself (at Stage One in Figure 4.7) would only be replaced where the energy required for the actual reinforcement manufacturing process was available (turning raw steel into reinforcing mesh). These further processing stages can often be excluded in a process analysis and result in *downstream truncation* of the system boundary (as discussed in Chapter 3).

It is evident from the figures in Table 4.4 that the input-output values for particular processes do not always provide an accurate representation of process-specific data. This is due to the number of limitations associated with input-output data, as outlined in Chapter 3. However, as is sometimes the case, the total combined process and input-output values of the energy inputs to the concrete production processes are slightly more closely aligned. This reflects the finding by Crawford (2005; 2008) that input-output values are less likely to be an accurate representation of the equivalent process value at the individual pathway or process level than at a whole material or building level.

One reason why the input-output value for concrete production for the residential building may be significantly higher than the process value given in Table 4.4 is that the input-output value for the concrete production (i.e. *Cement, lime and concrete slurry* into the *Residential building* sector) includes all instances of cement, lime, ready-mixed concrete and mortar purchased directly by the *Residential building* sector. As the process and input-output *total energy coefficients* for concrete production are fairly similar (3.45 and 3.05 GJ per m^3 respectively), this figure implies that the national average quantity of these products used within a residential building is higher than that used within the floor slab for this house. When assessing a whole building, for material inputs where process data is available, a substitution of a lower process value for a higher input-output value may lead to an underestimation of the environmental impacts associated with all instances of that material within the building. In reality, unless the practitioner can be certain that all instances of the requirements for that material have been quantified, only the proportion of the input-output value representing the quantified materials should be deducted from the input-output model. The other proportion should remain to cover those instances of the material not covered by process data. For example, here the requirement for ready-mixed concrete for the floor slab has been quantified, but other inputs from the *Cement, lime and concrete slurry* sector purchased directly by the *Residential building* sector, such as the cement used to make mortar and ready-mixed concrete used for other elements (such as for driveways, paths and other structural elements), have not. The difficulty lies in allocating input-output data to specific inputs due to the aggregation of different products within the same economic sector.

4.4.2.4 Using material coefficients

The use of material coefficients can help to simplify the data collection process and reduce the time and costs associated with collecting data. Process and input-output data can be integrated at the material level to provide hybrid material coefficients that cover the direct inputs and outputs of the manufacturing process and all of the inputs and outputs upstream of the material's manufacturing stage. For example, for concrete in Figure 4.7, this could cover the direct energy inputs associated with the concrete manufacturing process, all energy inputs for the processes required to produce other goods and services that provide a direct input into the concrete manufacturing process (such as cement and aggregate production), as well as all energy inputs for the processes required to produce these goods and services (such as mining of raw materials and transport at all stages) and so on, infinitely upstream in the supply chain. Similarly, material coefficients can be compiled for any other input or output to a product system (including water, greenhouse gases, pollutants and raw materials).

These coefficients are compiled using the same approach as used for the input-output-based hybrid assessment of the residential building in the previous section. For the relevant sector covering each material (e.g. the *Iron and steel* sector for steel and the *Sawmill products* sector for timber), available process data for specific production processes is substituted into the input-output model for the input-output pathways that they represent. Again, the price of the material is used to convert input-output values (in GJ of energy per A$1,000 of material, for example) to a total (energy) requirement per unit of material so that a direct substitution can be made. Equation 4.2 shows how an energy coefficient would be compiled using process and input-output data.

$$EC_m = PER_m + \left(TER_n - \sum_{i=1}^{I} (DER_i) \right) \times \frac{\$_m}{1,000} \qquad \text{Equation 4.2}$$

Where EC_m = hybrid energy coefficient of the basic material; PER_m = material process energy requirement per unit of material; TER_n = total energy requirement of the input-output sector n, representing the material, in GJ per A$1,000; DER_i = direct energy requirement of the input-output pathways representing the material production processes for which process data is available, in GJ per A$1,000; $\$_m$ = total price of the basic material.

Table 4.6 Hybrid energy coefficients for selected building materials

Material	Unit	Energy coefficient (GJ per unit)	Process data proportion (%)
25 MPa concrete	m^3	5.01	68.8
Aluminium	t	252.6	83.1
Clay bricks	m^2	0.56	71.4
Clear float glass (4 mm)	m^2	1.73	34.2
Fibre cement sheet (6 mm)	m^2	0.288	28.3
Fibreglass insulation (100 mm)	m^2	0.217	44.6
Hardwood	m^3	21.33	78.7
Plasterboard (10 mm)	m^2	0.207	16.8
Softwood	m^3	10.93	60.0
Steel	t	85.46	31.9
Steel decking	m^2	0.796	13.4
Toughened glass (6 mm)	m^2	3.66	24.3
Water-based paint	m^2	0.096	14.9
Wool carpet	m^2	0.741	29.3

Note: Energy coefficients were produced by Treloar and Crawford (2010) using an energy-based input-output model developed by Prof. Manfred Lenzen at the University of Sydney and Australian process data compiled by Grant (2002).

A pre-compiled database of material coefficients can then be used to assess the inputs from and outputs to nature for any building or infrastructure system. This can be extremely useful where a number of buildings or design variations that contain essentially the same materials but different quantities are being assessed. A list of hybrid energy coefficients for some main building materials is shown in Table 4.6, based on Australian process and input-output data for the late 1990s. A more detailed list is given in Table A.2, Appendix A. The proportion of these figures made up of process data gives an indication of their likely reliability.

Energy coefficients for materials are also available from a number of other sources. These include the *Inventory of Carbon and Energy* compiled by the University of Bath (University of Bath 2008) and the *Embodied Energy and CO$_2$ Coefficients for New Zealand Building Materials* report compiled by the Victoria University of Wellington (Alcorn 2003).

4.4.2.5 Avoiding sideways truncation of the system boundary

When using material coefficients to assess the inputs or outputs associated with the production of a building (or any other product), one remaining system boundary truncation problem still exists. Even if all materials at Stage One of the supply chain have been identified, other inputs exist at this stage (considered to be *sideways* of those other quantified Stage One inputs) that are not as easily quantified, such as those associated with the service sectors (financial, communications, advertising etc.) as shown in Figure 4.7 as *further sideways inputs*. By deducting the complete pathways from the input-output model of the *Residential building* sector that represent all of the quantified materials (for example, the complete supply chain for concrete from Figure 4.7, as this has been quantified using the concrete material coefficient), the *remainder* covering all otherwise unquantified processes can be deduced. This *remainder* can then be added to the figure calculated using the material coefficients to determine the total value of all inputs within the product system.

Equation 4.3 represents the calculation of the total energy required to construct a building (its embodied energy (EE)), including all of the energy associated with the processes upstream of the main construction process.

$$EE_b = \sum_{m=1}^{M}(EC_m \times Q_m) + \left(TER_n - \sum_{m=1}^{M}(TER_m) \right) \times \frac{\$_b}{1,000}$$ Equation 4.3

Where EE_b = total embodied energy of a building; EC_m = hybrid energy coefficient of material, m; Q_m = quantity of material, m; TER_n = total energy requirement of the building construction-related input-output sector, n, in GJ per A\$1,000; TER_m = total energy requirement of the input-output pathways representing the material production processes for which process data is available, in GJ per \$1,000; $\$_b$ = total cost of the building.

RESIDENTIAL BUILDING CASE STUDY – QUANTIFICATION OF ENERGY REQUIREMENTS

Embodied energy: Using input-output analysis, an initial indication of the total embodied energy of the house was calculated by multiplying the cost of the house (based on a rate of A\$650 per square metre of building area) by the *total energy requirement* of the Australian *Residential building* sector.

\$200,005 x 10.633 GJ per A\$1,000 = 2,127 GJ (6.9 GJ per m²)

Material quantities were then collected for individual building materials based on the bill of quantities for the house. These quantities were multiplied by their respective energy coefficient to determine the initial embodied energy of the house. A breakdown of the main materials used and calculation of their embodied energy is provided in Table A.3, Appendix A.

To fill the remaining data gaps for the minor and non-material-based processes, input-output data was used. The disaggregated input-output model of the Australian *Residential building* sector was used to identify the pathways covered by the material energy coefficients (Table A.4, Appendix A). The *total energy requirements* of these pathways (TER_m) were then subtracted from the *total energy requirement* of the *Residential building* sector (TER_n) to give the remainder to be added to the initial embodied energy figure (Equation 4.4). The calculation of the embodied energy of the house is shown in Table 4.7.

$$EE_b = \underset{a}{\sum_{m=1}^{M}(EC_m \times Q_m)} + \left(\underset{b}{TER_n} - \underset{c}{\sum_{m=1}^{M}(TER_m)} \right) \times \underset{e}{\frac{\$_b}{1,000}}$$ Equation 4.4

Table 4.7 Calculation of total embodied energy of residential building case study

Sum of initial embodied energy (GJ) (Table A.3)[a]	3,082
TER of *Residential building sector* (GJ/A$1,000)[b]	10.633
TER of I-O pathways covering initial embodied energy (GJ/ A$1,000) (Table A.4)[c]	6.074
Remainder (GJ/A$1,000)[(b-c)]	4.558[d]
House cost ($)[e]	200,005
Total embodied energy of the house (GJ)[(a+(d x e/1000))]	3,994

Note: Symbols a-e relate also to Equation 4.4.

This total embodied energy figure equates to 12.98 GJ per square metre of floor area. This figure is in line with other similar studies based on a hybrid LCI approach (for example, Fay *et al.* 2000).

Operational energy: Based on the gas and electricity bills collected for the house for an average year, the following (delivered) energy requirements were determined (Table 4.8). Primary energy values were then calculated based on primary energy factors for the fuels used (3.4 for electricity and 1.4 for natural gas for Victoria, Australia (Treloar 1998)).

Table 4.8 Delivered and primary operational energy consumption of residential building case study (GJ)

	Annual	*50 years*
Total delivered natural gas consumption	35.26	1,763
primary consumption[a]	49.36	2,468
Total delivered electricity consumption	11.56	578
primary consumption[b]	39.29	1,964
Total primary energy consumption[(a+b)]	88.65	4,432

4.4.3 Comparison of life cycle inventory approaches

Unless an input-output analysis is performed as part of the LCI phase, the practitioner often has no indication of the significance of those processes for which process data has not been collected. This can have serious repercussions for the reliability and completeness of the study results, any conclusions based on them and any actions taken in response to these conclusions.

As a way of comparing the various LCI approaches, Figure 4.8 provides an indication of the significance of the gaps inherent in each method. Based on the results of the residential building case study embodied energy assessment presented in this chapter, the graph shows that the input-output-based hybrid LCI approach clearly provides a higher embodied energy figure than any of the other LCI approaches commonly in use (26 to 278 per cent higher).

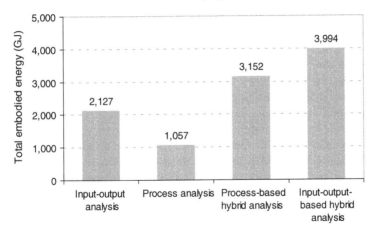

Figure 4.8 Comparison of residential building case study embodied energy based on different life cycle inventory methods

An input-output analysis provides a more comprehensive coverage of embodied energy requirements than a process analysis despite its limitations, as discussed earlier. This is the reason for the difference in embodied energy figures between these two approaches in Figure 4.8. The process-based hybrid analysis figure is based on the process analysis figure and includes input-output data upstream of the individually quantified materials, thus completing the system boundary for all processes upstream of these quantified materials. However, as discussed previously, this may exclude other minor materials and other goods and services required in the construction of a building. The input-output-based hybrid figure includes the energy inputs associated with the production or supply of all of these remaining materials, goods and services, thus completing the system boundary with further input-output data. In this case, these remaining energy inputs represent 21 per cent of the total embodied energy of the case study house.

Overall, the component of the input-output-based hybrid embodied energy figure represented by process data is 26.5 per cent. This means that had only process data been used to assess the embodied energy of this house, the energy requirements would have been underestimated by over 70 per cent.

4.4.4 Reuse and recycling

The reuse or recycling of construction materials can make a significant contribution to minimizing the environmental impacts associated with the built environment. Waste materials are created during almost every stage of the built environment life cycle, as a result of the manufacturing processes, construction, maintenance, refurbishment and eventual demolition. Depending very much on its quality and resource value, much of this waste is fed back into the supply chain to turn into new materials for use in new buildings. Quite often, materials are also recovered from other industries to make new building materials, such as the use of scrap steel from old cars to make structural steel members (scrap steel typically makes up at least 20 per cent of the raw materials used to make new steel). There is significant scope to redirect the millions of tonnes of waste materials that still go to landfill throughout the world every year into useable products. Decisions made by building designers and engineers on the choice of construction materials can have a significant impact on reducing this wastage, including specifying recycled and recyclable materials, as well as designing for reusability, durability and disassembly. This topic was discussed in detail in Chapter 2.

The processes involved in the reuse or recycling of waste materials must be understood and assessed as these too can be extremely resource-intensive, sometimes not much less so than the production of virgin materials. The reuse of existing materials (such as bricks or timber) may involve a limited amount of additional resources and the impacts avoided through their reuse can be quite significant compared to those associated with producing new materials. This may include a reduction in the impacts associated with resource depletion, waste going to landfill and the energy and water requirements and emissions associated with production (International Energy Agency 2004). Any additional impacts, such as those associated with transportation to site (if reused away from the original site) and making good of materials are often easily offset by these benefits. As the reuse of construction materials often involves very little re-processing of the materials, this process usually occurs between the demolition and construction, maintenance or refurbishment stages of the built environment life cycle (refer to Figure 4.1).

Recycling of construction materials typically involves some form of additional processing to convert old materials into new materials for either a similar use or a different use (e.g. turning scrap glass from bottles into glass fibre used to make fibreglass insulation). This may involve changing some of the inherent

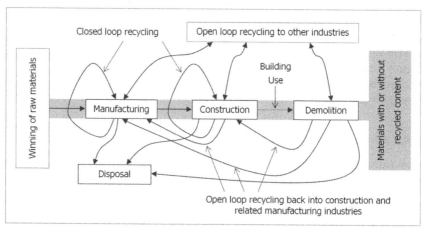

Source: Adapted from Treloar *et al.* 2003.

Figure 4.9 Flow of materials for recycling, reuse and disposal across the building life cycle

properties of the material. Materials for recycling can come from any stage of the life cycle and are often fed back into the manufacturing stage for use in new materials. However, as Figure 4.9 shows, there are also numerous other sources of materials, including open-loop recycling from other industries and closed-loop sources within existing manufacturing processes.

Depending on the extent of the manufacturing processes involved, significant additional resources may be required to reprocess these waste materials. For example, the recycling of glass into new glass requires around 75 per cent of the energy needed to produce glass from virgin materials (The EarthWorks Group 1990). On the other hand, for aluminium production, the conversion of alumina to aluminium is the most energy-intensive process involved (accounting for about two-thirds of the energy required to produce aluminium (EIA 2001)) and so recycling existing aluminium into new aluminium, which avoids the requirement for the alumina production and conversion processes, can result in an energy saving of up to 95 per cent (The EarthWorks Group 1990) compared to producing virgin aluminium from bauxite ore.

4.4.4.1 A credit for recyclability or reuse potential?

There is some belief that if a material can be reused or recycled after its initial use, the initial structure in which the material is used should be credited with the environmental benefits resulting from this potential reusability or recyclability. The problem with this view is that no guarantee can be made that these materials will be recovered for reuse or recycling at the end of their initial use, particularly due to the long life associated with many elements within the built environment. So, whilst the specification of materials or components of buildings and infrastructure that have the ability to be reused or recycled should be a priority to avoid many of the environmental impacts associated with their disposal and the use of virgin materials, it is also important to consider the implications of this not occurring, possibly due to changing priorities, technologies, expectations or additional knowledge gleaned over time. For these reasons, material specifiers should maintain a preference for materials that exhibit preferred environmental qualities such as reusability or recyclability, but any credit attributable to a reduced environmental impact resulting from this reusability or recyclability should be given to the structure in which they are reused (International Energy Agency 2004). For example, the environmental impacts resulting from the initial manufacture of steel beams should be attributed to the building or structure in which they are initially intended to be used. If these steel beams are reused in their existing state in a new building or structure at the end of their initial use, then impacts attributable to this new use would be limited to any additional resources or processes associated with the steel's transportation and installation in the new building or structure. If, however, the steel beams are recovered for recycling into a new product, then the impacts associated with the recovery and recycling processes should also be attributed to the new building in which the new product is used (Treloar *et al.* 2001).

The use of reused materials or materials with recycled content can sometimes increase the frequency of material replacement as durability can sometimes be lower for these types of materials. The decision to use these materials must be balanced against the different environmental priorities of the project, such as the avoidance of resource depletion or impacts from the processes associated with reuse.

4.4.5 Life cycle inventory studies

Often for buildings, and particularly where the study is limited to one or two environmental parameters (such as energy or emissions), the study may go straight from the LCI phase to interpretation, skipping the impact assessment. The nature of this limited scope of an LCI study (as opposed to an LCA study) means that a complex impact assessment may not be necessary in this situation. Where the goal of a study is to identify a limited range of resource inputs or outputs (for example, to reduce energy or raw material consumption) in order to target environmental improvement strategies, the relative quantities of these inputs or outputs may be sufficient to enable decisions to be made. An example of where an LCI study may be appropriate is for calculating the energy consumption across the various stages of the building's life cycle in order to reduce the life cycle energy requirements of the building. LCI findings can be used to identify the particular life cycle stage or building component representing the greatest energy requirement so that strategies such as optimizing embodied energy or operational energy efficiency through the selection of alternative materials, components or systems can be employed.

For studies where multiple environmental parameters are considered or a comparison between the environmental impacts of alternate products is being made, an impact assessment is typically essential to relate the findings from the LCI stage to impacts on the environment.

4.5 Life cycle impact assessment and interpretation

For the built environment, much of the LCI phase of an LCA is based on collecting objective data detailing the inputs and outputs associated with the construction or manufacture and operation of buildings, infrastructure systems and their associated components. The life cycle impact assessment phase (LCIA) involves the translation of these LCI results into numerical indicators that can be used to evaluate the environmental impacts attributable to the various inputs and outputs quantified.

Most of the steps involved in the LCIA phase are not nearly as specific to the built environment as those involved in the LCI phase. This phase uses the outputs of the LCI, which are typically in units of resources consumed or waste and pollutants released. The range of possible outputs is generally consistent across any product or service and any major variation would typically be due to variations in the characteristics of the inputs and outputs quantified for different products (such as the type of energy used and its source).

4.5.1 Selection of impact categories

Due to the complex nature of the many elements within the built environment, the impacts occurring throughout a building's life cycle cover a broad range of potential impact categories. Most of these are attributable to the manufacture and eventual disposal of materials, but also relate to the resources consumed during operation (such as energy and water). For example, material manufacturing processes consume raw materials, which leads to the depletion of natural resources; consume energy, which results in the depletion of energy reserves and release of greenhouse gas emissions; and produce pollutants and waste that lead to contamination of land, air and water, which in turn has an effect on human and environmental health. This complexity is one reason why building LCAs are often simplified by using a more streamlined LCA approach. In a streamlined LCA, the impact assessment will be somewhat less involved than for a full LCA. A streamlined LCA that chooses to assess only one environmental parameter may involve the consideration of only one or two different impact categories. Where multiple environmental parameters are being considered, the range of possible impact categories can increase significantly.

4.5.2 Classification

Where certain inputs or outputs contribute to more than one of the environmental impacts being considered, the practitioner must decide how this should be dealt with. Should they be assigned wholly or on a proportional basis to each impact category? For example, the materials used to construct a building require energy to produce, and this production (when based on fossil fuels) results in the release of emissions, such as nitrogen dioxide (NO_2). Nitrogen dioxide is classified as a greenhouse gas and is thus included within the *global warming* impact category in an LCA. It is also toxic to humans when inhaled and so can be classified within a *human toxicity* impact category. Should the environmental loadings associated with these materials (for example, the output of nitrogen dioxide) be assigned to both of these impact categories equally or should they be apportioned based on some assumption of their importance or other characteristic? This type of decision is one of the many that must be made by the LCA practitioner that can often be quite subjective and lead to considerable variability in the eventual findings.

4.5.3 Characterization

Once the LCI results have been classified into the relevant impact categories, they must be converted to the appropriate characterization units for those categories. This allows similar impacts resulting from various inputs, outputs or life cycle stages of the built environment to be grouped so that an overall impact can be ascertained (e.g. global warming potential from energy consumption and pollutant releases from the construction, operation and demolition of a building).

For example, where the impact category being assessed is *global warming*, any inputs of energy (possibly in a variety of energy units) will, first, need to be converted to quantities of the various greenhouse gases released as a result of their production, based on the fuel mix and technologies used. Where the various fuels used are known, emissions factors such as those listed in Table 4.9 can be used to convert energy units to emissions. As is quite often the case for complex items such as buildings, the quantity of the various fuels used in manufacturing the constituent materials and components is difficult to establish, particularly for the processes for which process data is not available. In this situation, it may be acceptable to use an average emissions factor to estimate the total emissions released from the combined fuel types, such as the figure of 60 kg CO_2-e per gigajoule of energy consumed, as used by Treloar (2000) for Australia.

The characterization factor is then used to convert the various quantities of greenhouse gases (e.g. kilograms of CO_2, CH_4 and N_2O) to a unit of equivalence based on the degree to which each input or output contributes to the specific environmental impact (e.g. global warming). In this case the characterization factor is the global warming potential (GWP) of the various gases, typically measured in carbon dioxide equivalents (CO_2-e). The global warming potential for the three main anthropogenic-based greenhouse gases over a 100-year timeframe are shown in Table 4.10.

Table 4.9 Emissions factors for combustion of some common fuel types

Fuel	Emissions factor (kg per GJ)		
	CO_2	CH_4	N_2O
Black coal	88.2	0.0014	0.00065
Brown coal	92.7	0.00048	0.0013
Natural gas	51.2	0.0048	0.000097
Wood	0.0	0.0038	0.0039
Gasoline	66.7	0.0095	0.00065
Diesel oil	69.2	0.0048	0.00065
Ethanol	0.0	0.0029	0.00065

Source: Department of Climate Change 2008.

Table 4.10 Global warming potential for common greenhouse gases relative to CO_2

Greenhouse gas	Global warming potential (CO_2-e)
Carbon dioxide (CO_2)	1
Methane (CH_4)	25
Nitrous oxide (N_2O)	298

Source: IPCC 2007b: 212.

Table 4.11 Example calculation of global warming category indicator results for 100 GJ consumption of natural gas

Energy use (GJ)		Emissions factor (kg/GJ)*		GWP^		Total (kg CO_2-e)
100	x	51.2 (CO_2)	x	1	=	5,120
100	x	0.0048 (CH_4)	x	25	=	12
100	x	0.000097 (N_2O)	x	298	=	2.9
			Total greenhouse gas emissions			5,134.9

*From Table 4.9, ^From Table 4.10.

The characterized LCI results for each input, output and life cycle stage can then be combined within each impact category by summing their values now that they are based on the same unit of measurement (as the example in Table 4.11 shows for the production of 100 GJ of natural gas).

4.5.4 Normalization, grouping and weighting

The characterized LCI results can then be normalized where necessary. For a whole building assessment, this is often useful on a per metre squared of floor area basis to enable a comparison across building design options. The category indicator results for each impact category are divided by the total floor area of the building, which then facilitates the comparison of environmental impacts between buildings of varying size or characteristics. The results from the LCIA for buildings may be grouped according to the impact categories assessed or by life cycle stage. This latter grouping can be beneficial for identifying which life cycle stage of the building is responsible for the greatest environmental impacts so that environmental improvement strategies can be targeted towards these areas. Weighting factors can then be used to rank the importance of the different impact categories. This weighting will often be based on the particular circumstances surrounding the significance of different issues in the specific geographic location where these impacts are likely to occur.

4.5.5 Study limitations

It is important that the limitations of the study results are clearly stated. These may be related to the availability of data, or lack thereof; the assessment methods used; characterization factors used; and judgements made about the importance of specific impacts. The influence of these limitations on the final study outcomes can be assessed using a sensitivity analysis.

RESIDENTIAL BUILDING CASE STUDY – IMPACT ASSESSMENT

As stated in the goal definition phase, the purpose of this case study is to quantify the global warming impacts from the construction and operation stages of a detached single-storey house.

Impact category:	Global warming
Category indicator:	Global warming potential (CO_2-e)
Characterization factors:	Embodied energy – 60 kg CO_2-e per GJ, Operational energy – as per Table 4.9.

The emission of greenhouse gases associated with the production and supply of energy embodied in the house and used for its operation were quantified based on the quantity of energy consumed, in primary terms.

Embodied energy emissions:
Due to the difficulties associated with determining the proportion of embodied energy supplied by the various fuel types within all of the processes involved in manufacturing and supplying the components of the house, an average emissions factor of 60 kg CO_2-e per GJ of energy has been used (Treloar 2000).

Embodied energy of house:	3,994 GJ (Table 4.7)
Emissions factor:	60 kg CO_2-e per GJ
Total embodied emissions:	239.6 t CO_2-e

Operational energy emissions:
The primary energy quantities of electricity and gas consumed for the house's operation (Table 4.8) were multiplied by the emissions factors for the three main greenhouse gases (Table 4.9) and then by their respective GWP (as per Table 4.10) to determine the total GWP of the operational energy requirements over 50 years.

Table 4.12 Calculation of operational energy-related emissions over 50 years for residential building case study

Fuel type	Quantity (GJ)	GHG	Emissions factor (kg/GJ)	GWP	Total GWP (kg CO_2-e)
Electricity	1,964	CO_2	92.7	1	182,063
		CH_4	0.00048	25	23.6
		N_2O	0.0013	298	760.9
Natural gas	2,468	CO_2	51.2	1	126,362
		CH_4	0.0048	25	296.2
		N_2O	0.000097	298	71.3
				Total	309,577

4.5.6 Interpretation

Once the LCI results have been classified into the appropriate impact categories, characterized in the appropriate units and normalized, grouped and weighted if necessary, the results can then be evaluated and conclusions

Table 4.13 Structuring building life cycle inventory data by life cycle stage, as a percentage

LCI input/ output	Raw material extraction	Manufacture	Construction	Operation	Maintenance & refurbishment	Disposal	Total
Energy	7.3	34	3.3	44	9.6	1.8	100
Water	6.1	32	1.5	48	11	1.4	100
Waste	1.1	5.1	9.8	0	19	65	100
CO_2	8.0	27	2.3	52	9.1	1.6	100

drawn as part of the interpretation phase of the LCA. Results evaluation efforts should be targeted towards those inputs, outputs, life cycle stages, assumptions and methods that represent the greatest contribution towards the study results. Inherently, any uncertainty in these aspects will also have the greatest potential for influencing the variability of the study findings. The data and information obtained from the study should be structured in a way that allows easy identification of the significant issues and drawing of conclusions. The LCI or LCIA data can be structured in a number of ways: by life cycle stages, individual processes, data types or sources or impact categories. Table 4.13 shows an example of some LCI data structured by building life cycle stage, represented as a percentage of the total inputs and outputs.

From this information the most significant contributors to the overall impacts can be identified. For example, in Table 4.13 the building operation stage contributes the greatest proportion of the overall energy requirements across the building life cycle (44 per cent) and the disposal stage generates the greatest proportion of waste over the life cycle (65 per cent). Once the most important aspects have been identified, variations to the data used or the assumptions made can then be assessed.

4.5.6.1 Evaluation of results

A sensitivity analysis can be used to assess how sensitive the study results are to changes in any of the data, methods or assumptions used. The degree of confidence that can be attributed to these results and any subsequent conclusions and recommendations can then be ascertained. Typically, a sensitivity analysis will be performed first on the aspects of the study where most uncertainty exists or on those aspects that represent the greatest contribution to the overall results, showing whether any of the decisions made during the study have had a significant influence on the final result.

The results of the study are compared to the results that would have been obtained had alternative data, methods or assumptions been used. By varying the data or assumptions by a set range (e.g. ±20 per cent), any significant

Table 4.14 Sensitivity analysis on life cycle inventory data source

	Operational energy
Original data source (GJ)	2,800
Alternative data source (GJ)	2,200
Deviation (GJ)	600
Deviation (%)	−21.4
Sensitivity (%)	21.4

changes in the results that may be influenced by the study uncertainties can be identified. Assuming that an alternative source of operational energy data was available for the example given in Table 4.13, Table 4.14 shows that the sensitivity of the results to the source of data chosen is significant, at up to 21.4 per cent.

Sensitivity analysis is particularly important in comparative studies, where even the slightest variations in data or assumptions used may change the results and the subsequent recommendations made based on the preferred option.

4.5.6.2 Conclusions, limitations and recommendations

The conclusions of the study will be informed by the results of the LCI, LCIA and interpretation phases and should address the goals of the study. The limitations of these conclusions should also be discussed, particularly in terms of those aspects that may have strongly influenced the findings, such as data availability, assumptions made or exclusions that were beyond the scope of the study. This phase of an LCA is extremely subjective, despite being based on predominately objective data. The practitioner's own interpretation of the results will direct any recommendations that are made and these may differ from person to person.

Recommendations will be based upon the intended goals of the study and use of the study findings and will depend on the type of study undertaken. These recommendations will provide the valuable information necessary to support the decision-making processes of the various stakeholders involved across the life cycle of the built environment. For example, comparative LCAs of materials or building components will be useful to architects and building designers for specifying environmentally preferable building elements. A possible recommendation may be the use of one material over another, due to the potential for lower environmental impacts across its life cycle.

LCAs of specific materials or products will be useful to material and product manufacturers for promoting the environmental credentials of these items and informing product improvement. Recommendations may include a list of the most significant issues where environmental improvement can be achieved, possible strategies for improving the efficiency of particular manufacturing processes and suggestions for possible material or product modifications.

Whole building LCAs will be useful for building designers, managers and owners to inform building design, operation and upgrade strategies. Recommendations may include informing the design process by identifying areas where the greatest reductions in environmental impacts can be achieved or which design solution may result in the lowest impact; areas to which improvement strategies should be targeted; and particular strategies for improving environmental performance.

RESIDENTIAL BUILDING CASE STUDY – INTERPRETATION

Identification of significant issues
Table 4.15 indicates that over the 50-year life cycle for the house the two life cycle stages considered represent similar proportions of the total energy requirements and GWP-related impacts.

Table 4.15 Significance of energy inputs and category results over 50-year life cycle for residential building case study

Life cycle stage	LCI input (%)	Category result (GWP) (%)
Embodied energy	47.4	43.6
Operational energy	52.6	56.4
– electricity	(44.3)	(59.1)
– gas	(55.7)	(40.9)
Total	100	100

Sensitivity analysis
The sensitivity of the results to the uncertainties associated with the data used is shown in Table 4.16. This shows that the uncertainty associated with the embodied energy data used is significant, particularly as a result of the use of national average input-output data. Overall, the sensitivity of the results to the use of this data is quite considerable (30 per cent). The GWP impacts from embodied energy consumption are likely to range from 168 to 311 t CO_2-e.

Table 4.16 Sensitivity analysis for data uncertainty on category results for residential building case study

	Embodied energy	Operational energy	Total
GWP (t CO_2-e)	239.6	309.6	549.2
Deviation (%)	±30	0	
GWP range (t CO_2-e)	168 – 311	309.6	477.6 – 620.6
Sensitivity (%)	30	0	23

Conclusions, limitations and recommendations
The study shows that over a 50-year life for the building and life cycle stages analyzed, the total energy consumed is 8,425 GJ (27.4 GJ/m²). This figure is subject to a possible variation of ±14.2 per cent based on an embodied energy uncertainty rate of 30 per cent, accounting for the issues associated with the data used for embodied energy analysis. The embodied energy and operational energy represent close to equivalent proportions of the total life cycle energy

requirements. The GWP of the operational energy consumed is slightly higher than that for the energy embodied in the house. The GWP associated with the total energy use is 477.6 – 620.6 t CO_2-e.

Energy (and GWP) associated with maintenance, refurbishment and eventual disposal of the building materials have not been included, and these too may be significant.

As this house has already been constructed, recommendations from a design point of view may only be useful as knowledge for future building designs. In this case recommendations may focus on strategies for reducing operational energy consumption further, such as behavioural changes to the use of heating and cooling systems, blinds, opening windows and so forth. If any of the active systems in the house were due to be replaced, then selecting more efficient replacements may also be recommended. However, replacing existing items of equipment (such as heating, cooling and hot water systems, refrigerators and other appliances) prior to the end of their physical life is usually not warranted, considering the associated additional embodied impacts from their manufacture.

4.5.7 What to do with the results?

How the results are used will be determined by the intended purpose of the study. A further assessment may be necessary if the goal of the LCA was to provide information to direct environmental improvement strategies, in order to confirm that alterations to the building or product will actually result in a net environment improvement. Take the situation where the impacts associated with a particular HVAC system for use in a commercial office building have been analyzed through an LCA. The study has identified that the impacts associated with the energy consumed by the system far outweigh any other impacts associated with the system over its life (including those resulting from its manufacture). An alternative is chosen that is known to significantly reduce the energy required for its operation; however, unknowingly, the energy required to manufacture it is significantly higher than the original solution. If a decision is made to choose an alternative system based on reducing these operational impacts alone, without assessing the life cycle impacts of the alternative, then the potential exists for there to be a net increase in environmental impacts, despite what may be the best intentions.

4.6 Life cycle assessment tools for the built environment

There is a range of tools available worldwide for performing anything from simple environmental assessments up to full LCAs on buildings and other components of the built environment. These tools range from simple spreadsheet calculations to more sophisticated modelling tools that provide environmental information to assist in the identification of the most significant environmental issues and comparisons of alternative designs or products. These may take anywhere from less than an hour to up to several weeks to perform a single assessment.

The main goal of these LCA tools is to simplify and organize the assessment process through the selection and analysis of product and environmental data, the assessment of environmental impacts and reporting of study findings. Many of these tools have come about due to the lack of a clear and simple approach for accessing and using LCA information effectively. Whilst these tools need to be sophisticated enough to consider a potentially large and complex range of environmental issues, they must also be simple and understood easily enough that they can be used by the various stakeholders for which they are intended. For example, architects and building designers are constantly faced with a myriad of demands, not least the often restrictive time and financial resources available for providing a finished built outcome. More often than not, environmental attributes of a project are the first to be excluded when project funds and timelines are constrained. Providing a simple and quick to use, yet comprehensive, LCA tool for designers is essential if life cycle thinking is to be fully integrated across the built environment. Helping to make LCA simpler, more accessible and more easily integrated into current practices can also assist in the accelerated uptake of LCA by all of the parties involved in procuring the various elements of the built environment.

LCA tools containing pre-established environmental data for specific or generic materials and processes can greatly reduce the time involved in conducting an LCA, as the time involved in the data collection process can be substantially reduced. However, one of the limiting factors to the use of pre-established environmental data is the availability of geographic and temporally specific data representative of the product under assessment. Any tool is only as good as the data contained within it and while some tools allow the user to import or up-date existing data that may be more relevant from a geographic or temporal perspective, the relevance and accuracy of any results can be affected by the quality of that data.

The outputs of LCA tools can be used for the same range of uses for which the outputs of a conventional LCA are used, such as marketing, product development, decision making or product improvement purposes. The most sophisticated LCA tools are often able to assess a broad range of products and processes, well beyond just those associated with the built environment. There also exists LCA tools specific to the built environment, which are especially used to assess and improve the environmental performance of buildings.

4.6.1 Built-environment-specific life cycle assessment tools

These tools are intended to be used to assess buildings and sometimes other elements that form part of the built environment. They generally focus on energy and energy-related emissions, but also consider the depletion of natural resources, production of waste, water consumption, release of pollutants and human health impacts. Two such examples are the *ATHENA® Impact Estimator for Buildings* and *Building for Environmental and Economic Sustainability (BEES)* tools.

4.6.1.1 ATHENA® Impact Estimator for Buildings by the Athena Sustainable Materials Institute

The ATHENA® Impact Estimator is a whole building, life cycle environmental decision support tool for use by building designers, product specifiers and policy analysts at the conceptual design stage of a project. The tool focuses on assessing industrial, institutional, office and residential type buildings. It indicates the environmental implications of different combinations of materials or design choices, allowing the user to consider the trade-offs among the various environmental impacts. Up to five design scenarios can be compared across a range of selected environmental parameters. Multiple existing building designs can be compared or strategies for improving the environmental performance of a single design can be identified. Every stage of the building life cycle is considered, although at different levels of sophistication. This includes the extraction of raw materials, material manufacturing, transportation, on-site construction, operational energy consumption, on-going maintenance, repair and replacement, demolition and disposal. Direct and indirect impacts associated with the energy consumed during the operational stage of the building are based on user-inputted annual operating energy by fuel type. This considers the global warming impacts associated with the production of the energy, as well as the related emissions to air, water and land over the life cycle of the building. The embodied impacts are calculated based on a comprehensive database of regionally specific environmental attributes for each of the constituent materials. These attributes include the quantity of raw materials consumed, energy use, water use, emissions to air and water and solid waste produced in order to manufacture a specified quantity of these materials. The user defines the building assemblies for each element of the building in terms of their geometry and material make-up or composition. A bill of quantities is compiled as the user selects an assembly, which is then combined with the environmental attributes of each material to determine the total environmental impact of the building.

The impacts associated with maintenance and repair factor in the frequency of these activities for each material, the quantity of new materials required, the energy usage for these activities, on-site waste and the transportation of these components. The impacts associated with the replacement of materials that occurs over the life of the building are also included. These impacts are based on the type of building, the typical life expectancy (service-life) of each material, the quantity of waste material and how these are disposed. Energy required for demolition and the transportation to landfill for those materials assumed to be landfilled at the end of the building's life is also included. The regional variation in energy use and transportation is also factored in.

The combined quantities of inputs of energy, water and raw materials and outputs of emissions and waste for the whole design are categorized as per Table 4.17 and converted to the appropriate characterization units.

Table 4.17 Impact categories covered by the ATHENA® Impact Estimator tool

Impact category	Characterization factor
Global warming	Carbon dioxide (CO_2) equivalents
Acidification	Hydrogen ion (H+) equivalents
Human health respiratory effects	Particulate matter equivalents
Eutrophication	Nitrogen (N) equivalents
Photochemical smog	Nitrogen oxides (NOx) equivalents
Ozone depletion	Trichlorofluoromethane (CFC-11) equivalents
Primary energy	Gigajoules

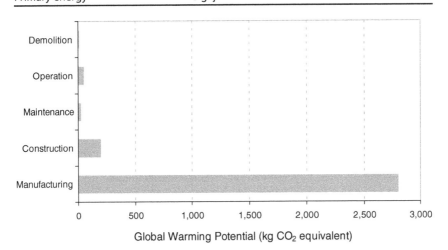

Source: Adapted from ATHENA 2009.

Figure 4.10 Example results from the ATHENA® Impact Estimator tool, by life cycle stage

Results obtained from the assessment are presented graphically by life cycle stage for each of the impact categories, as shown for the global warming category in Figure 4.10. The comparison of up to five alternative designs can be made either on a whole building level or on a per square metre of floor area basis.

One of the limitations of the tool is that the data contained within it and some of the assumptions made are only relevant to Canada and parts of the United States. Also, the LCI data for materials and components contained within the tool cannot be modified by the user, which also limits the ability for the tool to be adapted by the user for use in other regions of the world.

4.6.1.2 Building for Environmental and Economic Sustainability (BEES) tool by NIST

The Building for Environmental and Economic Sustainability tool, developed by the United States National Institute of Standards and Technology (NIST), assesses the environmental performance of building materials and products for commercial and residential buildings. Whilst based on the LCA framework, this tool does not allow an analysis at the whole building level.

The BEES tool uses quantitative data that is contained within the tool to assess the environmental performance of individual building materials and products across the impact categories outlined in Table 4.18. Similar products or materials can easily be compared across different environmental parameters or life cycle stages (raw material acquisition, manufacture, transportation, installation, use, recycling and waste management).

Results can be viewed by impact category, showing the impact of each category on the overall environmental performance (Figure 4.11) or by life cycle stage for the combined or any one of the impact categories (Figure 4.12). Impact scores can be combined by weighting each impact category by its relative importance to overall environmental performance.

Table 4.18 Impact categories covered by the BEES tool

Impact category	Characterization factor
Global warming	Carbon dioxide (CO_2) equivalents
Acidification	Hydrogen ion (H+) equivalents
Eutrophication	Nitrogen (N) equivalents
Fossil fuel depletion	Megajoules
Indoor air quality	Total volatile organic compound (VOC) emissions
Habitat alteration	Threatened and endangered species
Water intake	Litres
Criteria air pollutants	MicroDALY equivalents
Photochemical smog	Nitrogen oxides (NOx) equivalents
Ecological toxicity	Dichlorophenoxy-acetic acid (2,4-D) equivalents
Ozone depletion	Trichlorofluoromethane (CFC-11) equivalents
Human health	Toluene equivalents

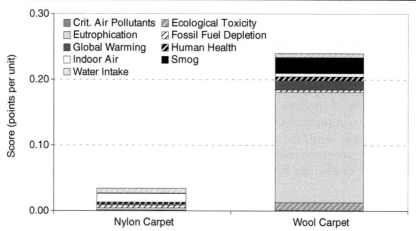

Source: Adapted from NIST 2007.

Figure 4.11 Comparing the environmental performance of alternative options using BEES, by impact category

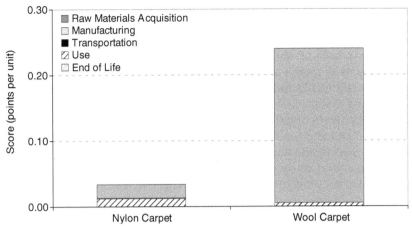

Source: Adapted from NIST 2007.

Figure 4.12 Comparing the environmental performance of alternative options using BEES, by life cycle stage

BEES also measures the life cycle economic performance of building products, considering initial capital costs and those associated with operation, replacement, maintenance, repair and disposal. The life cycle environmental impacts can be combined with the economic performance, based on a user-defined weighting factor to weight the relative importance of the environmental and economic components, to give an overall performance score.

4.6.1.3 Integrated design tools

Many building-specific LCA tools lack the ability for the user to visualize the building that is being assessed. More recent developments to these tools include the integration of LCA with computer-aided design (CAD) tools. By defining the characteristics of the individual elements and materials within the architectural CAD drawings for a building, these tools can generate a bill of quantities for the whole building and, using an inbuilt database of environmental impacts associated with specific materials, calculate the overall environmental impact of the building. Elements of the design can then be changed and re-assessed. One of the major advantages of these tools is the potential time savings for conducting an LCA. Examples of currently available CAD-based LCA tools include LCADesign (Ecquate 2009) and Ecotect (Autodesk 2010).

4.6.2 General life cycle assessment tools

These tools are not specifically designed to assess only products or elements within the built environment, but nevertheless do often focus on the assessment of the individual materials that are regularly used in constructing buildings and infrastructure systems. They also often focus on the assessment of manufacturing processes, packaging materials and smaller items or components, such as appliances and equipment. The number of processes covered by these tools is generally much broader than those contained within a building-related LCA tool; however, it is often not possible to use these tools for assessing a whole building.

There is a range of general LCA tools available worldwide, with the most commonly used and sophisticated of these being those such as SimaPro (PRé Consultants 2010), the Boustead Model (BCL 2007) and GaBi (PE International 2010). These tools all cover a wide range of impact categories and life cycle stages and often allow comparisons to be made between alternatives. A more detailed list of available LCA software tools is published by the European Commission (2010). Despite the individual benefits that each tool offers, and their respective limitations, it is also important to be aware of other issues that may provide certain constraints on obtaining accurate and relevant findings. These are particularly related to the data contained or used within these tools, especially its geographic relevance, age and source.

4.6.3 Limitations of existing life cycle assessment tools

All of the aforementioned tools can be useful for the particular purpose for which they are intended, provided that the user understands their potential limitations (for example, shifting of impacts to other life cycle stages that may not have been assessed). Even in this situation, the results can be useful if the intended goals of the study have been addressed.

Data reliability and completeness is the greatest concern in the use of LCA tools, as it is with LCA in general. The pre-established data contained within most LCA tools has been collected based on a process analysis framework. Some, such as SimaPro (PRé Consultants 2010) or the EIOLCA tool (Carnegie Mellon University 2002), may also contain input-output data, but no tools currently exist that provide an input-output-based hybrid framework to enable a comprehensive LCI analysis to be undertaken. Without the systemically complete product system boundary that an LCI based on input-output data provides, the reliability and usefulness of subsequent phases of an LCA can be significantly compromised. For this reason, one of the biggest issues with current LCA tools is akin to that for LCA in general: that there will always be a lack of reliability or a degree of incompleteness associated with the software outputs.

Some of these tools allow the user to input their own data for selected processes, but this can significantly extend the time needed to conduct an assessment. Also, inconsistencies in data collection methods or system coverage can have considerable consequences for the reliability of the study findings.

Most existing LCA tools contain data or parameters that restrict the tool's use to a particular geographic or regional location. Using data that is not representative of the product being analyzed can be problematic as impacts can differ substantially, even for similar products made in different countries.

This may be due to different fuel mixes, production processes, transportation distances and the source of raw materials. For some tools, it is possible to link to additional databases that are more representative of locations, processes or other characteristics of the product being analyzed. For example, SimaPro allows the user to import any database in a variety of formats. To date, databases covering production in a range of geographic locations, such as Japan, Europe and the United States, have been included through a multitude of databases, including Ecoinvent, BUWAL 250, The Franklin US LCI library, The ETH-ESU 96 library and Japanese, United States, Dutch and Danish input-output tables.

4.7 Summary

A life cycle approach to the assessment and minimization of environmental impacts for any product or process is essential in order to ensure that environmental improvement strategies will result in net benefits to the environment. This chapter has described the application of life cycle assessment for quantifying the environmental impacts associated with the built environment. A case study example of a residential building has been used to demonstrate the various assessment phases, including goal and scope definition, life cycle inventory analysis, impact assessment and interpretation.

A streamlined approach to LCA is often considered appropriate for buildings and other complex products within the built environment due to the many thousands of potential inputs and outputs that would need to be modelled even for one environmental parameter. The highly energy-intensive nature of the built environment, all the way from resource extraction through to its use, further warrants the energy-focused streamlined approach commonly used to assess many elements of the built environment.

This chapter has placed particular emphasis on the input-output-based hybrid LCI approach. This is seen as the most relevant approach for quantifying the environmental impacts of the built environment, providing a systemically complete assessment of the associated inputs and outputs of any product or process within an entire economy. Using this method, it is possible to conduct a much more comprehensive assessment of the environmental impacts resulting from the construction of any element of the built environment than is possible with previous process-based approaches.

A number of the software tools available for conducting an LCA in the built environment have also been described. These range from built-environment-specific tools to more generic tools used to assess a much broader range of products and processes. Whilst these tools attempt to simplify and organize the LCA process, the user must be aware of some of the limitations associated with their use, particularly those relating to the use of the data that is often contained within them. It is also important that the usefulness and applicability of any tool used for an LCA study is understood, as geographic and scope limitations (such as limited life cycle stages, impact categories or system boundaries) of some tools may reduce the relevance of any findings.

The following chapter provides a number of worked case study examples of LCA applied to various products and elements within the built environment. These cases show how LCA can, whilst not recommending any particular solution, assist in the environmental decision-making process.

5 Case studies

Examples of life cycle assessment in the built environment

The built environment is made up of a vast range of elements, all of which serve to support human existence but which also have a considerable impact on the environment. By using LCA to assess the extent of these impacts, it is possible then to identify and implement strategies that will help to alleviate some of the strain that the built environment poses to the natural environment, to which our very existence on Earth is intrinsically linked.

The LCA approach described in the previous chapter has been used here to perform various levels of LCA on a range of case studies reflecting various elements of the built environment. Whilst the range of case studies presented by no means provides an exhaustive view of all of the possible components of the built environment, it does provide a broad cross-section of some of the key elements on which human societies are heavily reliant. This includes three building-related case studies, from a streamlined study of a single whole building to comparative studies of two alternative structural construction approaches for a commercial building and two alternative external wall assemblies for a residential building; a comparative analysis of a variety of road types and two types of railway sleepers, both part of the broad transport networks and infrastructure that are now essential across the ever-expanding cities of the world; and two of the most commonly used technologies for producing clean renewable energy, namely a large-scale wind turbine and a domestic solar photovoltaic system.

Whilst each of these case studies is based on a streamlined LCA approach, mainly focused on energy consumption and its associated environmental impacts, this is enough to demonstrate the various phases involved in an LCA, the potential limitations, necessary decisions, and the possible recommendations that might be made based on the findings. These case studies also cover a number of environmental parameters, impact categories and approaches that may be used to evaluate the assessment results. Although only a limited number of environmental parameters and impact categories are considered in each of these case studies, the purpose here is not to provide a full LCA of each example. The objective is to demonstrate some of the key principles and issues associated with conducting an LCA, as well as some of the challenges that exist for improving the environmental performance of the built environment. These cases show how LCA can assist in the environmental decision-making process, whilst not necessarily making recommendations on any particular solution.

5.1 Building case studies

The complex nature of buildings means that there are hundreds, if not thousands, of decisions that must be made regarding their design, construction, operation, maintenance and eventual demolition. There is a broad range of possible design solutions, ranging from the selection of materials, to the building components and other systems that might be used within a building. This, as mentioned previously, is one of the key reasons why conducting a full LCA on a building project can be extremely time consuming and often not possible within the time and resource constraints of many building projects.

The growing demand for buildings with improved environmental performance means that possible environmental design solutions must be assessed to ensure that they provide improved environmental outcomes. LCA is an important tool for conducting such an assessment. This assessment may be conducted as part of the material or product development process, with the findings used to market a material or product's environmental credentials to building designers and specifiers, or at the building design stage to compare the environmental credentials of the possible choices of materials or building components being considered. The latter comparison is more necessary where the functional unit of assessment relates to a particular building design rather than a basic unit of material or a single product.

5.1.1 Case study 1: Selecting a building structural system – steel or reinforced concrete?

The purpose of this case study is to demonstrate how LCA can be used to select between competing or alternative construction materials or systems at the whole building level. It also highlights some of the decisions that may need to be made in this type of comparison and why a life cycle approach to this decision is absolutely essential.

Scope: The study quantifies the energy inputs associated with two variations to the construction of a commercial office building in order to determine the alternative with the lowest embodied energy and thus associated environmental impacts. These variations relate to the structural elements of the building and include a steel-framed construction and concrete-framed construction approach. All building elements – other than the beams, columns and slabs – were considered to be identical for both alternatives.

Impact category: Energy use
Category indicator: Gigajoules of energy

Functional unit: The functional unit for the assessment was the construction of a 75,570 m², 50-storey commercial building with external dimensions of

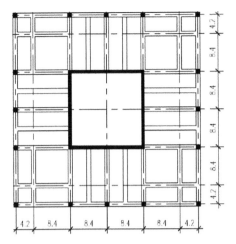

Figure 5.1 Plan view and image of commercial building

approximately 42 × 42 m² (Figure 5.1). The energy required to construct the building (in gigajoules) was assessed based on both the steel and concrete-framed construction methods using an identical floor plan. For the concrete-framed alternative, 40 MPa steel-reinforced concrete was used for the footings and columns, with 30 MPa steel-reinforced concrete used for the beams, slabs and staircases. The steel-framed building consisted of steel columns and beams, 40 MPa steel-reinforced concrete for the footings and 30 MPa steel-reinforced concrete slabs. The external façade for both construction types consisted of aluminium-framed double-glazed panels.

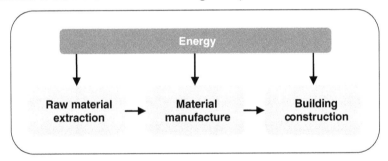

Figure 5.2 Life cycle stages considered for steel and concrete building life cycle assessment

System boundary: Only the energy inputs associated with the initial building construction were assessed, excluding energy required for on-going component replacement, end-of-life refurbishment, disposal and possible reuse and recycling of materials. All energy requirements associated with the actual construction process and all supporting processes and services upstream of this were included as per Figure 5.2 (such as, but not limited to, raw material extraction, transportation and manufacturing of materials and the provision of capital equipment). Where process data was not available, input-output data was used to fill any gaps in the construction system boundary (for example, for building services and internal fitout and finishes). The choice of structural system is unlikely to have any significant impact on the thermal performance and thus the operational energy requirements and associated environmental impact of a building and so this stage was not included in the analysis.

Figure 5.3 shows the individual material production and building construction processes for which process data was obtained. All remaining material and minor goods and services requirements were quantified using input-output data.

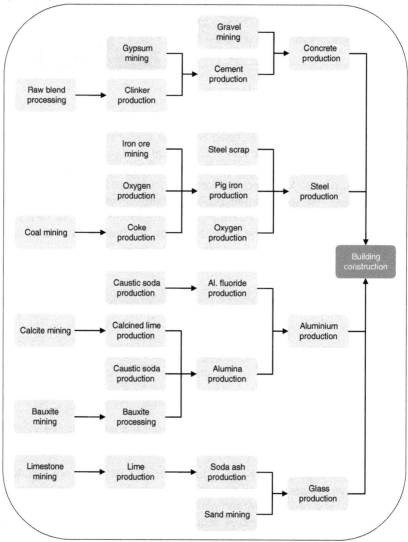

Figure 5.3 Processes included in process-based system boundary for steel and concrete building life cycle assessment

5.1.1.1 Life cycle inventory of steel and concrete-framed building construction

The building was analyzed to determine the quantities of materials that were able to be quantified (concrete and steel structural elements and façade glazing elements), for both steel and concrete structural variations (Tables B.1 and B.2, Appendix B). The energy requirements associated with the construction of the two building variations were calculated by multiplying these material quantities by their respective hybrid material energy coefficient (as listed in Table A.2, Appendix A). The resultant value represents all process-based energy requirements for the materials quantified, as well as all energy requirements associated with the provision of goods and services upstream of and associated with these materials (avoiding the common upstream truncation errors associated with a process analysis). The calculation of the initial embodied energy figure for the two building variations is shown in Table B.1 and B.2 of Appendix B.

Any remaining data gaps (those sideways and downstream of the main materials and processes) were then filled with input-output data in accordance with the input-output-based hybrid LCI approach outlined in Chapter 4. From the energy-based input-output model developed by Lenzen and Lundie (2002), the pathways representing the individual materials quantified for the two building variations (as per Figure 5.3) were identified within the *Other construction* sector and their *total energy requirements* (as per Table 5.1)

Table 5.1 Total energy requirement of input-output pathways covered by process data collected for steel and concrete building life cycle assessment

Sector	Element	Input-output pathway (materials covered)	TER (GJ/A$1000)
Other construction	Structure	*Cement, lime and concrete slurry* (footings, slabs, columns) (beams – concrete building only)	0.7914
		Iron and steel → Structural metal products (reinforcement, columns) (beams – steel building only)	0.2385
	Façade	*Glass and glass products* (glazing)	0.1061
		Basic non-ferrous metal and products (extruded aluminium framing)	0.0374
		Total	1.1734

TER = Total energy requirement. Note: *Iron and steel → Structural metal products* represents a demand for goods or services (and thus energy) from the *Iron and steel* sector by the *Structural metal products* sector (i.e. the steel used to make reinforcement). A limitation of input-output analysis is that it utilizes national average data, and as both building types fall within the *Other construction* sector and both contain the same, albeit different quantities of the same materials, the pathways (and thus TERs) for which process data was collected are identical for both construction types.

subtracted from the *total energy requirement* of the sector (9.9798 GJ/ A$1,000). The remaining energy inputs (those not subtracted from the *total energy requirement* of the sector) complete the system boundary for any previously excluded inputs sideways or downstream of the main materials.

The sum of these remaining energy inputs in GJ per A$1,000 of sector output was then related to the specific building by multiplying it by the total construction cost for both building variations to give the *remainder* value for each variation (Table 5.2). This figure was then added to the initial embodied energy figure calculated in the previous step to determine the total embodied energy of both the steel and concrete-framed buildings (Table 5.3).

Table 5.2 Calculation of remainder to fill sideways and downstream embodied energy data gaps for steel and concrete building

Sector	Structure	Sector TER (GJ/A$1000)[a]	Cost (A$)[b]	Sum TER of related paths (GJ/A$1000)[c]	Remainder (GJ/building) ((a–c) x b/1000)
Other construction	Steel	9.9798	292,002,480	1.1734	2,571,491
	Concrete	9.9798	275,452,650	1.1734	2,425,746

[c] From Table 5.1.

Table 5.3 Embodied energy calculation of steel and concrete building

	Embodied energy (GJ/building)	
	Steel	Concrete
Process data for quantified building materials[a]	412,516	321,326
Input-output data used to fill *upstream* gaps for building materials[b]	662,407	402,889
Initial embodied energy [(a+b)]	1,074,923[c]	724,215[c]
Input-output remainder (to fill *sideways* and *downstream* gaps)[(Table 5.2)]	2,571,491[d]	2,425,746[d]
Total [(c+d)]	3,646,414[e]	3,149,961[e]
Proportion of process data [(a/e)]	11.3%	10.2%

5.1.1.2 Interpretation – evaluation of results

The assessment of the total energy inputs required to construct the steel and concrete-framed buildings showed a 16 per cent increase in the energy required for the steel-framed approach compared to the concrete-framed alternative. Assuming potential errors in the LCI data obtained for all of the processes associated with the construction of the buildings, the sensitivity of the findings to variations in this data can be ascertained. Based on an error range of ±40 for the energy data used, Table 5.4 shows the possible effect of the data used on the comparison between the total embodied energy for both steel and concrete-framed buildings.

This sensitivity analysis shows that the potential errors associated with the data used for the study may have a significant impact on the findings, with a sensitivity of 40 per cent. This level of uncertainty, common particularly when using input-output data, may in the worst-case scenario actually reverse the main findings of the study so that the steel-framed building results in lower energy requirements than the concrete building. However, this would only be the case where the energy requirements associated with the construction of the steel and concrete buildings were significantly higher and lower than the national average (as represented by the input-output data), respectively.

Whilst the environmental impacts resulting from the initial construction of a building may represent a significant component of a building's life cycle impacts, these should generally not be considered in isolation. This is particularly important where two or more alternative construction approaches, materials or systems are being compared and impacts across other stages of the building life cycle may have a considerable influence on any decision to choose one approach over another. For example, a design solution for the steel-framed commercial building that allows for easy disassembly and reuse of the main structural steel components at the end of the building's life may actually warrant a higher initial impact if these longer-term benefits are seen to outweigh the higher shorter-term impacts. This solution may reduce longer-term impacts in two ways: first, by minimizing the quantity of waste materials sent to landfill and, second, by reducing the quantity of virgin materials needed in the subsequent building.

Table 5.4 Sensitivity analysis for energy data uncertainty on comparison of embodied energy for steel and concrete building

	Steel	Concrete	Difference
Original embodied energy (GJ)	3,646,414	3,149,961	496,459
Deviation (%)	±40	±40	
Embodied energy range (GJ)	2,187,848 – 5,104,980	1,889,977 – 4,409,945	3,215,003
Sensitivity (%)	40	40	

Of course, as with any LCA study, these findings may also vary depending on a range of other assumptions made, including variations to the manufacturing technologies used, source of raw materials, location of production and installation. Also, if the fuel mixes used are identical for both steel and concrete construction, then it might be acceptable to assume that the construction method with the lowest embodied energy value (concrete in this case) will have the lower environmental impact resulting from this energy consumption. However, if the fuel mixes used in the manufacture of the different materials were to differ, then the resulting environmental impacts may also vary considerably. This may be the case where less greenhouse-emissions-intensive fuel types (such as renewables) have been used during the material extraction, production and construction processes.

This case study has demonstrated the importance of taking a life cycle approach to the comparison between construction materials or systems in order to select the environmentally preferred alternative. The impacts associated with certain life cycle stages or the way in which end-of-life materials are treated, for instance, can have a significant impact on the outcome of an LCA study on competing materials or construction approaches.

5.1.2 Case study 2: Commercial office building

The purpose of this case study is to demonstrate how LCA can be used to identify from which element of a building's construction the greatest environmental impacts may occur. With this information, strategies for reducing these impacts can then be better targeted in order to address those impacts that are found to be most significant.

Scope: This study quantifies the life cycle energy requirements associated with a commercial office building in order to prioritize strategies for reducing its energy-related environmental impacts.

> *Impact category*:   Global warming
> *Category indicator*:   Global warming potential (CO_2-e)

Functional unit: The functional unit for the assessment was a three-storey 11,600 m^2 commercial office building occupied for a period of 50 years (Figure 5.4). The main materials used in the building are listed in Table 5.5. The total quantity of energy consumed to construct, operate, maintain, refurbish and demolish the building was determined in gigajoules and then converted to global warming potential (GWP) to reflect the potential energy-related impact across the building's predicted life of 50 years.

System boundary: All of the energy inputs associated with the initial building construction, operation, on-going component replacement, intermediate refurbishment and eventual disposal of materials have been determined (Figure 5.5).

The elements included in the assessment of the construction-related impacts for which physical material quantities were obtained included the main

Figure 5.4 Image of commercial office building

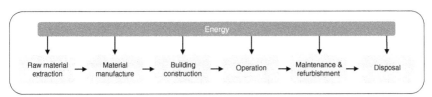

Figure 5.5 Life cycle stages considered for commercial office building life cycle assessment

Table 5.5 Materials used in commercial office building construction

Element	Materials
Substructure	200 mm thick reinforced concrete slab
Structure	Painted steel columns and steel reinforced concrete slabs
Staircases	Reinforced concrete staircases with painted steel frame and handrails
Roof	Steel roof frame and decking, aluminium-framed toughened glass glazed roof
External walls	Painted pre-cast concrete and aluminium cladding panels
Windows	Frameless toughened glass solar double-glazed windows, aluminium-framed toughened glass curtain walls, louvers, screens and windows
Internal walls	Single skin brickwork, 10 mm plasterboard-clad steel and 200 mm thick Hebel blocks
Floor finishes	50 mm insitu terrazzo, carpet tiles, marble, vinyl and ceramic tiles
Wall finishes	Painted 10 mm plasterboard, timber wall panelling, ceramic tiles
Ceiling finishes	Painted 13 mm plasterboard and 9 mm fibre cement sheet

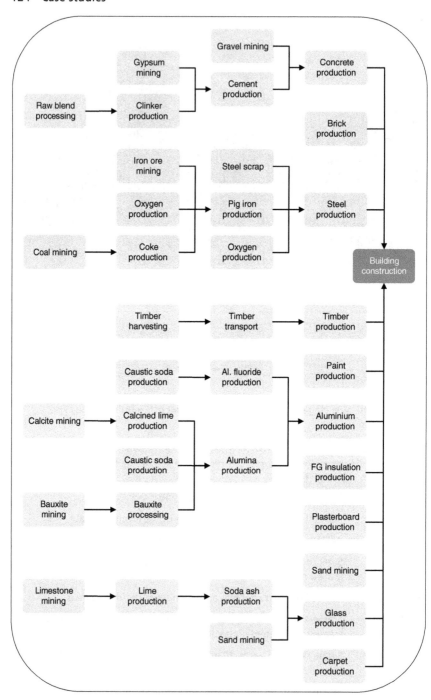

Figure 5.6 Processes included in process-based system boundary for commercial office
      building life cycle assessment

structure, floors, internal and external walls and finishes. The building services, such as HVAC equipment, lighting and electrical wiring, were accounted for with input-output data. Figure 5.6 shows the individual material production and building construction processes for which process data was obtained.

5.1.2.1 Life cycle inventory of commercial office building

BUILDING CONSTRUCTION

The energy consumption associated with the building's initial construction (including all of the energy required for the supporting processes upstream of the main construction process) was determined using the hybrid material energy coefficients and the input-output-based hybrid LCI approach to ensure that the complete system boundary relating to the building's construction was covered. The quantities of materials contained within the building were determined based on an analysis of the individual building components (Table B.3, Appendix B). These material quantities were multiplied by their respective hybrid material energy coefficient (as listed in Table A.2, Appendix A) to determine the process-based energy requirements for the materials quantified, as well as all energy requirements associated with the provision of goods and services upstream of and associated with these materials. The calculation of the initial embodied energy figure for the building is shown in Table B.3, Appendix B.

As per the input-output-based hybrid LCI approach outlined in Chapter 4, any remaining data gaps (those sideways and downstream of the main materials and processes) were then filled with input-output data. From the energy-based input-output model developed by Lenzen and Lundie (2002), the pathways representing the individual materials quantified for the building (as per Figure 5.6) were identified within the *Other construction* sector and their *total energy requirements* (as per Table B.4, Appendix B) subtracted from the *total energy requirement* of the sector. The remaining energy inputs (those not subtracted from the *total energy requirement* of the sector) complete the system boundary for any previously excluded inputs sideways or downstream of the main materials. The sum of these remaining energy inputs in GJ per A$1,000 of sector output was then related to the specific building by multiplying it by the total construction cost to give the *remainder* value for the building (Table 5.6), which was then added to the initial embodied energy

Table 5.6 Calculation of remainder to fill sideways and downstream embodied energy data gaps for commercial office building

Sector	Sector TER (GJ/A$1000)[a]	Cost (A$)[b]	Sum TER of related paths (GJ/A$1000)[c]	Remainder (GJ/building) ((a−c) x b/1000)
Other construction	9.97983	23,200,000	1.9323	186,703

[c] From Table B.4.

Table 5.7 Embodied energy calculation of commercial office building

	Embodied energy (GJ/building)
Process data for quantified building materials [a]	68,738
Input-output data used to fill *upstream* gaps for building materials [b]	125,869
Initial embodied energy [(a+b)]	194,607[c]
Input-output remainder (to fill *sideways* and *downstream* gaps) [(Table 5.6)]	186,703[d]
Total [(c+d)]	381,310[e]
Proportion of process data [(a/e)]	18%

figure calculated in the previous step to determine the total energy required to construct the building (Table 5.7).

BUILDING OPERATION

The energy consumed in operating the building was obtained from the energy bills, averaged over several years of consumption. The total energy consumed was found to be 6,466 GJ per annum. This energy is sourced from both brown coal-fired electricity and natural gas, which make up 40 and 60 per cent of the total energy consumed, respectively. Whilst in this case the operational energy consumption figure was taken from the energy bills for an existing building, this information may not always be available, especially for buildings that have not yet been constructed. In this situation, it may be necessary to predict likely energy usage based on prior knowledge or experience or by using building energy simulation tools.

BUILDING MAINTENANCE, REFURBISHMENT AND DISPOSAL

Due to a lack of energy data related to the maintenance, refurbishment and disposal of buildings, several assumptions have been made to determine an approximate value for these components. Combined energy requirements associated with maintenance and refurbishment of the building over its life have been assumed to be equivalent to 1 per cent of the energy required for the building's initial construction, per annum. This equates to 190,655 gigajoules of energy over the anticipated life of the building. The energy associated with disassembly or demolition and subsequent disposal of materials at the end of a building's life has been shown by Crowther (1999) to represent less than 1 per cent of the energy required across the building life cycle. To account for this energy, a figure equivalent to 1 per cent of the life cycle energy requirements of the building has been included in the life cycle energy figure for the building. This equates to 12,962 gigajoules of energy.

5.1.2.2 Impact assessment – life cycle greenhouse gas emissions

The life cycle energy requirements of the building were converted to GWP impacts using the results from the LCI and characterization factors. A characterization factor of 60 kg CO_2-e per GJ was used to convert the energy required for initial construction, on-going maintenance, refurbishment and disposal to GWP impacts. The GWP associated with the energy required for the building's operation was calculated by converting the delivered operational energy value into primary energy terms using a factor of 3.4 for the electricity component and 1.4 for the natural gas component. Characterization or emissions factors for brown coal-fired electricity and natural gas were then used to convert the primary energy consumption figures into emissions terms. GWP factors for the three main greenhouse gases (CO_2, CH_4 and N_2O) were then used to convert the emission figures into CO_2-e terms. The life cycle emissions associated with the building were then able to be determined, as in Table 5.8.

Table 5.8 Life cycle greenhouse gas emissions of commercial office building

Life cycle stage	Primary energy (GJ)	GHG	Emissions factor (kg per GJ)^	GWP*	Total emissions (t CO_2-e)
Construction	381,310	CO_2-e	60		22,879
Operation (50 years)					
Electricity	439,688	CO_2	92.7	1	40,759
"		CH_4	0.01	25	110
"		N_2O	0.4	298	52,411
Natural gas	271,572	CO_2	51.2	1	13,904
"		CH_4	0.1	25	679
"		N_2O	0.03	298	2,428
Maint. and refurbishment	190,655	CO_2-e	60		11,439
Demolition and disposal	12,962	CO_2-e	60		778
			Total life cycle emissions		145,387

^*Source:* Department of Climate Change 2008: 202.
*Global Warming Potential, *Source:* IPCC 2007b: 212.

5.1.2.3 Interpretation – evaluation of results

The assessment of the life cycle greenhouse gas emissions associated with the commercial office building showed that a total of 145,387 tonnes of CO_2-e emissions are produced over the life of the building. The greatest proportion of these emissions (76 per cent) can be attributed to the energy required for the building's operation (for heating, cooling, lighting and other services) (Figure 5.7). This should therefore be where most environmental improvement efforts are focused in order to make the greatest contribution to reducing the life cycle energy-related environmental impacts of the building.

Assuming potential errors in the LCI data obtained for all of the stages associated with the building's life cycle, the sensitivity of the findings to variations in this data can be ascertained. Based on an error range of ±40 for the data used to calculate the construction, maintenance, refurbishment, demolition and disposal energy requirements, Table 5.9 shows the possible effect of the data used on the total life cycle emissions of the building. Operational energy requirements have not been included here as these were based on actual measured consumption figures.

This sensitivity analysis shows that the potential errors associated with the data used for the study may have a significant impact on the findings, with a sensitivity of 40 per cent. The contribution of the non-operational energy-related emissions may range from 14 to 34 per cent of the total life cycle emissions of the building.

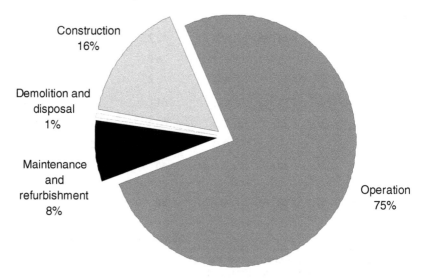

Figure 5.7 Life cycle greenhouse gas emissions of commercial office building, by life cycle stage

Table 5.9 Sensitivity analysis for energy data uncertainty on impact category results for commercial office building

Life cycle stage	Total emissions produced (t CO_2-e)*
Construction[a]	22,879
Maintenance and refurbishment[b]	11,439
Demolition and disposal[c]	778
Total (non-operational) [(a+b+c)]	35,096
Deviation (%)	±40
Range	21,058 – 49,134
Sensitivity (%)	40

*Total emissions calculated by multiplying energy requirement by an emissions factor of 60 kg CO_2-e/GJ.

5.1.3 Case study 3: Construction assemblies

By the time that a building design has progressed to a stage where its environmental performance can be assessed at a whole building level using LCA, the opportunities for improvement have been diminished, as many decisions on systems, planning and materials have already been made. History shows that bringing about any major changes to these decisions at this point is difficult. Hence, building designers need reliable environmental information as early as possible during the design process to ensure that environmental outcomes can be maximized. Whilst it is not possible to assess the environmental performance of a whole building before it has even been designed, having environmental performance information for individual building materials and components readily available can greatly assist building designers to make more informed choices early on.

It is generally not necessary to replicate the assessment of materials or components for every project as existing data can more readily be used, thus streamlining the assessment and environmental decision-making processes. Environmental performance data for common construction components and materials can be pre-assessed and compiled into a database or tool to streamline the material selection decisions made during the building design process.

A number of tools or databases that facilitate the assessment of the environmental performance of construction materials or components currently exist. These are often based on an LCA of the constituent assembly materials, covering a range of life cycle stages, from initial embodied impacts to the impacts associated with operation, maintenance and eventual disposal. For example, the ATHENA® EcoCalculator for Assemblies (ATHENA 2007) enables an assessment of the global warming potential, fossil fuel depletion, embodied energy, and the pollution to air and water associated with a wide combination of building assemblies, covering all stages from raw material extraction through to final disposal, but excluding operational impacts.

This case study demonstrates how this level of information can be beneficial to building designers when selecting environmentally preferred building components. For example, by providing the energy-related global warming impacts associated with two (or more) external wall assemblies for a residential building, a designer can compare and select the preferred option that minimizes GWP impacts from that component of the building across the building life cycle.

Scope: The energy inputs associated with the construction (including all supporting inputs), maintenance and refurbishment of two residential building external wall assemblies over a 50-year period were quantified. These energy inputs were converted to carbon dioxide equivalents (CO_2-e) to indicate the GWP impacts of each assembly.

Impact category:	Global warming
Category indicator:	Global warming potential (CO_2-e)

Functional unit: A 1 m^2 area of external wall was assessed for each assembly type over a lifetime of 50 years. The details for the assemblies are shown in Table 5.10.

System boundary: The energy inputs associated with raw material extraction, material manufacture, construction, maintenance (such as re-painting) and refurbishment (replacement of materials) were assessed, as per Figure 5.8. The impacts and benefits associated with eventual reuse or recycling of materials were excluded.

Table 5.10 Characteristics of construction assemblies

Assembly	Description
Brick veneer	• Standard clay brick external cladding • Aluminium reflective foil insulation • 90 x 45 softwood internal framing • R2.0 fibreglass batt insulation • 10 mm plasterboard internal lining • Water-based paint (internal finish)
Timber weatherboard	• 25 mm hardwood weatherboard external cladding • Aluminium reflective foil insulation • 90 x 45 softwood internal framing • R2.0 fibreglass batt insulation • 10 mm plasterboard internal lining • Water-based paint (external & internal finish)

Figure 5.9 shows the individual production processes for which process data was obtained for the two assemblies.

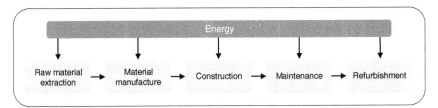

Figure 5.8 Life cycle stages considered for construction assembly life cycle assessment

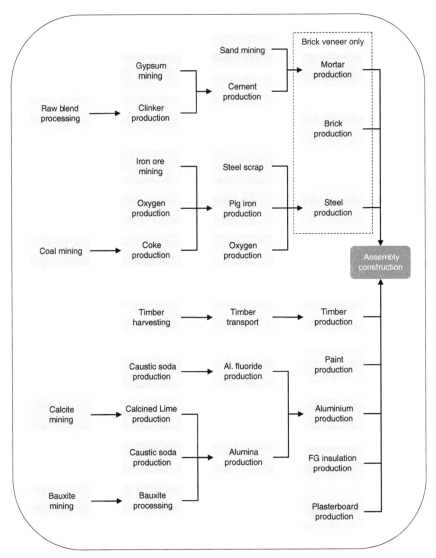

Figure 5.9 Processes included in process-based system boundary for construction assembly life cycle assessment

5.1.3.1 Life cycle inventory of assembly construction, maintenance and refurbishment

ASSEMBLY CONSTRUCTION

The material quantities contained within each of the assemblies were determined based on common construction techniques. The energy requirements associated with the assemblies' initial construction were calculated by multiplying these material quantities by their respective hybrid material energy coefficient (Table A.2, Appendix A). The resultant value represents all process-based energy requirements for the materials quantified, as well as all energy requirements associated with the provision of goods and services upstream of these materials (avoiding the common upstream truncation errors associated with process analysis). The calculation of the initial embodied energy figure for the construction of the assemblies is shown in Table 5.11 and Table 5.12.

Input-output data was then used to fill any remaining data gaps (those sideways and downstream of the main materials) in accordance with the input-output-based hybrid LCI approach outlined in Chapter 4. From the energy-based input-output model developed by Lenzen and Lundie (2002), the pathways representing the individual materials quantified for the assemblies (as per Figure 5.9) were identified within the *Residential building* sector (as per Tables B.5 and B.6 of Appendix B for the brick veneer and timber weatherboard assemblies, respectively).

Table 5.11 Material quantities and initial embodied energy calculation for brick veneer assembly construction

Component	Material	Material quantity	Unit	Embodied energy coefficient (GJ/unit)	Embodied energy (GJ/m^2)
External cladding	Clay bricks	0.85	m^2	0.56	0.476
	Mortar	0.0314	t	2.00	0.063
Wall ties	Steel	0.00005	t	85.46	0.005
Insulation	Aluminium reflective foil	1	m^2	0.14	0.137
	R2.0 fibreglass insulation	0.869	m^2	0.18	0.159
Framing	Softwood	0.0118	m^3	10.90	0.129
Internal lining	Plasterboard (10 mm)	1	m^2	0.21	0.207
Internal finish	Water-based paint	1	m^2	0.096	0.096
				Total	1.270

Note: Figures may not sum due to rounding.

Table 5.12 Material quantities and initial embodied energy calculation for timber weatherboard assembly construction

Component	Material	Material quantity	Unit	Embodied energy coefficient (GJ/unit)	Embodied energy (GJ/m²)
External finish	Water-based paint	1	m²	0.096	0.096
External cladding	Hardwood timber	0.025	m³	21.3	0.53
Insulation	Aluminium reflective foil	1	m²	0.14	0.14
	R2.0 fibreglass insulation	0.869	m²	0.18	0.16
Framing	Softwood	0.0118	m³	10.9	0.13
Internal lining	Plasterboard (10 mm)	1	m²	0.21	0.21
Internal finish	Water-based paint	1	m²	0.096	0.096
				Total	1.36

The combined *total energy requirement* of these pathways was subtracted from the *total energy requirement* of the *Residential building* sector. The remaining energy inputs (those not subtracted from the *total energy requirement*) complete the system boundary for any previously excluded inputs sideways or downstream of the main materials. The sum of these remaining energy inputs was then converted to energy terms by multiplying it by the construction cost of each assembly to give the *remainder* value for the assemblies (Table 5.13), which was then added to the initial embodied energy figure calculated in the previous step to determine the total embodied energy associated with the initial construction of the assemblies (Table 5.14).

Table 5.13 Calculation of remainder to fill sideways and downstream embodied energy data gaps for construction assemblies

Assembly	Sector TER (GJ/A$1000)[a]	Cost (A$/m²)[b]	Sum TER of related paths (GJ/A$1000)[c]	Remainder (GJ/m² assembly) ((a−c) x b/1000)
Brick veneer	10.633	117	3.173	0.87
Timber weatherboard	10.633	117.05	1.084	1.12

[c] From Tables B5 and B6.

Table 5.14 Embodied energy calculation of construction assemblies

	Embodied energy (GJ/m² assembly)
Brick veneer assembly:	
Initial embodied energy [Table 5.11]	1.27[a]
Input-output remainder (to fill *sideways* and *downstream* gaps) [Table 5.13]	0.87[b]
Total [(a+b)]	2.14
Weatherboard assembly:	
Initial embodied energy [Table 5.12]	1.36[c]
Input-output remainder (to fill *sideways* and *downstream* gaps) [Table 5.13]	1.12[d]
Total [(c+d)]	2.48

ASSEMBLY MAINTENANCE AND REFURBISHMENT

The energy associated with the maintenance and replacement of materials (also known as the recurring embodied energy) during the assumed 50-year life cycle of the building in which the assemblies would be part was calculated based on assumed maintenance requirements and replacement rates for the constituent materials, as per Table 5.15.

As for the initial embodied energy calculation, the energy embodied in replacement materials was calculated by multiplying the material quantities by the hybrid material coefficients and filling the data gaps associated with energy inputs sideways and downstream of these materials with input-output data, as per the hybrid LCI approach. Table 5.16 and Table 5.17 show the calculation of the recurrent embodied energy for the two assemblies.

Table 5.15 Replacement rates for common building materials

Material	*Useful life*	*Material*	*Useful life*
Aluminium reflective foil	50[^]	Steel wall ties	50[^]
Clay bricks	50[^]	Hardwood (external)	25[#]
Fibreglass insulation	50[^]	Softwood (internal)	50[#]
Mortar	50[^]	Water-based paint	10[*]
Plasterboard	30[*]		

Note: Based on 50 year building life. *Source:* *Treloar *et al.* (2000a), #FWPA (2007), ^assumed.

Table 5.16 Calculation of recurrent embodied energy of brick veneer assembly

	Initial embodied energy (GJ/m²)	No. of replacements	Recurrent embodied energy (GJ/m²)
Internal lining (plasterboard)	0.21	1	0.21[a]
Internal finish (paint)	0.096	4	0.38[b]
Recurrent embodied energy [(a+b)]			0.59[c]
Input-output data for replacement materials:			
Total energy requirement*			0.47[d]
Inputs covering process data^			0.29[e]
Input-output remainder (to fill *sideways* and *downstream* gaps) [(d–e)]			0.18[f]
Total recurrent embodied energy of material replacement [(c+f)]			0.77

*Based on the cost of the replacement materials (A$43.78) and the TER of the *Residential building* sector (10.633 GJ/A$1,000). ^This value equates to the sum of the TERs of the related pathways (including those processes considered not to be associated with the replacement materials, e.g. concrete and steel production) (6.60 GJ/A$1,000) multiplied by the cost of the replacement materials.

Table 5.17 Calculation of recurrent embodied energy of timber weatherboard assembly

	Initial embodied energy (GJ/m²)	No. of replacements	Recurrent embodied energy (GJ/m²)
External cladding (hardwood timber)	0.53	1	0.53[a]
Internal lining (plasterboard)	0.21	1	0.21[b]
Internal and external finish (paint)	0.19	4	0.77[c]
Recurrent embodied energy [(a+b+c)]			1.51[d]
Input-output data for replacement materials:			
Total energy requirement*			1.205[e]
Inputs covering process data^			0.738[f]
Input-output remainder (to fill *sideways* and *downstream* gaps) [(e–f)]			0.467[g]
Total recurrent embodied energy of material replacement [(d+g)]			1.97

*Based on the cost of the replacement materials (A$113.33) and the TER of the *Residential building* sector (10.633 GJ/A$1,000). ^This value equates to the sum of the TERs of the related pathways (including those processes considered not to be associated with the replacement materials, e.g. iron and steel production) (6.513 GJ/A$1,000) multiplied by the cost of the replacement materials.

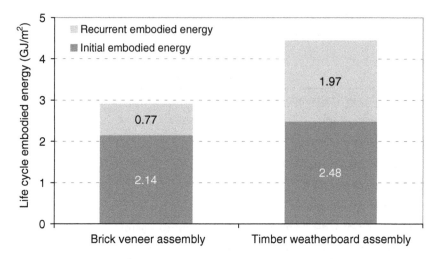

Figure 5.10 Life cycle embodied energy requirements of brick veneer and timber weatherboard assemblies

These initial and recurrent embodied energy figures were then summed to determine the total life cycle embodied energy requirements for the construction, maintenance and refurbishment of both assemblies, as depicted in Figure 5.10.

5.1.3.2 Impact assessment – life cycle greenhouse gas emissions

The life cycle embodied energy requirements of the two assemblies were converted to GWP impacts using the results from the LCI and the characterization factor of 60 kg CO_2-e per GJ (Table 5.18).

Table 5.18 Life cycle greenhouse gas emissions of construction assemblies, per m^2 of assembly

Assembly	Life cycle embodied energy (GJ)	GHG	Emissions factor (kg per GJ)	Total emissions produced (t CO_2-e)
Brick veneer	2.911	CO_2-e	60	0.175
Timber weatherboard	4.453	CO_2-e	60	0.267

5.1.3.3 Interpretation – evaluation of results

The findings from the impact assessment phase show that over the 50-year life cycle considered the timber weatherboard assembly results in 53 per cent greater GWP impacts than the brick veneer alternative. This impact results not only from the greater quantity of energy required for the initial manufacturing

Table 5.19 Significance of category results for construction assemblies over 50-year life cycle

Life cycle stage	Category result (GWP) (%)	
	Brick veneer	Timber weatherboard
Initial embodied energy	73.6	55.6
Recurrent embodied energy	26.4	44.4
Total life cycle energy	100	100

and construction stage, but also from the energy requirements and thus GWP impacts associated with maintenance and replacement during the considered life cycle. For the timber weatherboard assembly, these recurrent impacts represent a 44 per cent contribution towards the total life cycle GWP impact (Table 5.19).

This case study shows the importance of being able to assess impacts that occur across various stages of the building life cycle. In this case, the impacts associated with on-going maintenance and materials replacement of the assemblies are quite significant. Employing particular design strategies that address only those impacts resulting from initial construction (such as using materials with a lower embodied energy) may be counter-productive. A more holistic approach to environmental design that attempts to minimize life cycle impacts (such as through the specification of more durable materials that may even involve a greater investment of embodied energy to begin with) is often a better solution.

The most obvious and simplest recommendation that stems from these findings may be that the brick veneer option would be the most preferred from the point of minimizing greenhouse gas emissions. However, there is also much more that can be gleaned from these findings. For example, areas for further reductions in energy consumption and emissions in both options, as well as the identification of materials or components that may be able to be substituted with more durable or less emissions-intensive alternatives, can be identified.

The thermal performance of building assemblies, particularly for external elements, will influence the quantity of energy required to run heating and cooling systems to keep internal temperatures to an acceptable level. This energy and the associated environmental impacts from its production should also be considered in a broader assessment of the life cycle impacts of particular assembly choices. However, the exact quantity of energy consumed will be dependent on a range of factors beyond the characteristics of the assemblies themselves, including the location, size, design and orientation of the building, as well as occupant behaviour.

Energy requirements associated with building assemblies or any other building component are obviously only one of a number of the resource inputs and emission and waste outputs that may be attributed to them. A broader

LCA will consider the other environmental issues that may arise from the use of particular building assemblies and this information is useful for building designers to make the most informed environmental choices. The weighted importance of environmental parameters, the durability and recyclability of materials, local environmental conditions and the source of raw materials may also have an impact on the preferred choice with the lowest long-term impact on the environment.

5.2 Transport infrastructure case studies

As the world's population continues to be drawn towards large cities, transport systems are becoming more and more essential to aid their movement and improve or maintain the efficiency of these cities. As a city's population grows, so does the need to build new and upgrade existing transport networks and all of the associated infrastructure that goes with this. As for other elements of the built environment, such as buildings, there often exists a number of competing or alternative options for the design of many of the elements of these transport networks. Many of the decisions that must be made in the design of new or upgrade of existing road and rail networks relate to the materials used in their construction. Whilst the financial and functional aspects of the various options have long been key factors in the decision-making process, as environmental issues gain much greater prominence on a global level, the environmental impacts associated with the possible options are becoming just as important. The following two case studies provide examples of how LCA can be – and has been – used within transport planning to select the most environmentally preferred option for railway sleeper and road construction materials.

5.2.1 Case study 4: Railway sleepers

The purpose of this case study is to demonstrate how LCA can be used to select between competing or alternative materials based on particular environmental priorities. It also highlights some of the decisions that may need to be made in weighting environmental parameters or impact categories. Also, the importance of defining an appropriate functional unit and considering the durability and anticipated service-life of competing materials is demonstrated.

Scope: The study quantifies the energy and water inputs and greenhouse gas emissions associated with the production of two alternative railway sleeper (also known as cross-ties) materials. These materials include reinforced concrete and hardwood timber (River red gum or Eucalyptus camaldulensis). The properties of the sleepers are shown in Table 5.20.

Functional unit: The functional unit for the assessment was a 1 km length of track. Reinforced concrete sleepers will generally be expected to last longer than hardwood timber sleepers. In this situation, time is an important consideration when defining the functional unit. It is important that the

Table 5.20 Railway sleeper properties

Reinforced concrete sleeper:	
Concrete weight	285 kg per sleeper
Concrete strength	50 MPa, based on Australian Standard 1085.14 (2003)
Steel reinforcement	2,700 x 8 dia x 4 per sleeper
Fastening details	10 kg per sleeper
No. of sleepers per km track	1,400 (714 mm spacing)
Timber sleeper:	
Species	River red gum, Eucalyptus camaldulensis
Air dry density	900 kg/m^3
Dimensions	2,850 x 125 x 250
Weight	80.2 kg per sleeper
No. of sleepers per km track	1,460 (685 mm spacing)

varying periods of replacement are factored in through a consideration of related impacts over an extended period of time. In this case, a 100-year life cycle was considered.

System boundary: The energy and water inputs and greenhouse gas emissions associated with the initial and replacement sleeper production were assessed, excluding inputs and outputs associated with maintenance, end-of-life disposal and possible reuse and recycling of materials. All energy and water requirements and greenhouse gas emissions associated with the actual timber and concrete sleeper production processes and all supporting processes and services upstream of this were included, as per Figure 5.11 (such as, but not limited to, raw material extraction, transportation and manufacturing of materials and the provision of capital equipment). Steel fastenings used to connect the rails to the sleepers were also included. Ballast and rails were excluded from the analysis as these were considered to be identical for both sleeper types.

The assessment of the emissions associated with timber sleeper production and use included not only the direct and indirect emissions resulting from the energy required to produce the sleepers and their fastenings, but also the non-energy-related emissions resulting from the burning or decay of roundwood conversion waste from the harvesting and production processes, the decay of sleepers over time and a credit for the carbon stored in the sleepers.

Figure 5.12 shows the individual sleeper material production processes for which process data was obtained.

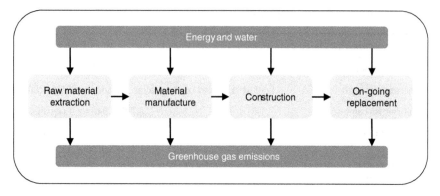

Figure 5.11 Life cycle stages, inputs and outputs considered for railway sleeper life cycle assessment

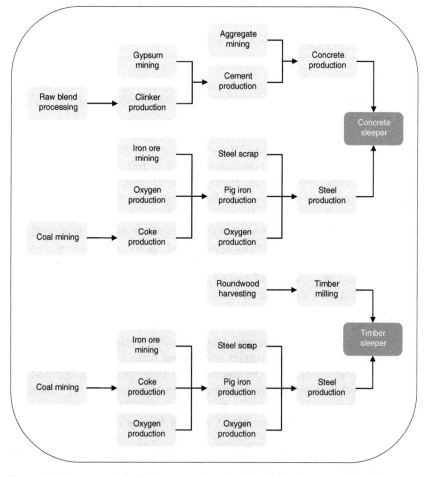

Figure 5.12 Processes included in process-based system boundary for reinforced concrete and timber sleeper life cycle assessment

Table 5.21 Impact categories and indicators for railway sleeper life cycle assessment

Impact category	Category indicator
Energy use	Gigajoules of energy (GJ)
Water use	Kilolitres of water (kL)
Global warming	GWP (t CO_2-e)

Three impact categories were selected for the LCA of the sleeper options, addressing what were considered to be the most important environmental impacts resulting from sleeper production and use, namely global warming and the depletion of water and energy resources (Table 5.21).

5.2.1.1 Life cycle inventory of reinforced concrete and timber sleeper production

The energy and water requirements and greenhouse gas emissions associated with the production of the two sleeper types were calculated by multiplying the quantities of the constituent sleeper materials by their hybrid material coefficient. As per the input-output-based hybrid LCI approach outlined in Chapter 4, input-output data was then used to fill any remaining data gaps (those sideways and downstream of the main materials and processes) to calculate the total initial energy and water requirements and greenhouse gas emissions associated with the production of the two sleeper types. This figure, per sleeper, was then converted to a per kilometre basis based on the number of sleepers required for this length of track.

While the service-life of the two sleeper types can vary considerably depending on usage frequency, loads and environmental conditions, the service-life that has been assumed for the sleepers was 50 and 30 years for the reinforced concrete and timber sleepers, respectively. This consideration of service-life is essential to ensure that any comparative assessment of competing or alternative options factors in the life cycle benefits of a prolonged service-life and greater durability. Recurrent inputs of energy and water and greenhouse gas emissions produced as a result of sleeper replacement over a 100-year life cycle were calculated based on the number of replacements required for each sleeper type (one for concrete and three for timber). Life cycle energy and water inputs and emission outputs associated with both sleeper types were then calculated by combining the initial and total recurring inputs and outputs during a 100-year period (Table 5.22).

Table 5.22 Life cycle energy, water and emissions of reinforced concrete and timber
sleepers

Parameter	Sleeper type	Initial input/ output	Recurrent input/output	Life cycle input/output
Energy (GJ/km)	Reinforced concrete	4,335	4,335	8,670
	Timber	6,658	18,388	25,046
Water (kL/km)	Reinforced concrete	8,814	8,814	17,627
	Timber	9,589	26,482	36,071
Emissions (t CO_2-e/km)	Reinforced concrete	328	328	656
	Timber	812	2,244	3,481*

*Includes 425 t CO_2-e resulting from the decay of sleepers used over 100 years.
Note: Figures may not sum due to rounding.

5.2.1.2 Impact assessment

As the values obtained from the LCI phase for each of the impact categories
were already in the required category indicator units, it was unnecessary to
undertake a characterization of the LCI results. However, each impact category
was weighted based on its perceived impact relative to the magnitude of the
impacts caused by the other two impact categories (Table 5.23).

The calculated weighting results show that energy consumption is the most
important factor for the manufacture of timber sleepers. However, for
concrete sleepers, water consumption is the most important factor. Overall,
across all three impact categories considered, the impact of timber sleepers is
around 155 per cent greater than that of the reinforced concrete alternative.

Table 5.23 Weighting of impact categories and life cycle inventory results for railway
sleeper life cycle assessment

Impact category	Life cycle input/output per km of sleepers		Weighting factor	Weighting result	
	Concrete	Timber		Concrete	Timber
Energy use (GJ)	8,670	25,046	0.3	2,601	7,514
Water use (kL)	17,627	36,071	0.2	3,525	7,214
Global warming (t CO_2-e)	656	3,481	0.5	328	1,740
			Total	6,454	16,468

5.2.1.3 Interpretation – evaluation of results

This analysis showed that, regardless of the perceived importance of each of the impact categories considered, the reinforced concrete sleepers result in lower environmental impacts than timber sleepers. Despite this finding, further improvements in the environmental performance associated with the production of both sleeper types are possible. This may involve improving the resource efficiency of the manufacturing process or sourcing alternative materials as a substitute for the resource-intensive materials currently being used, such as cement used in the manufacture of concrete sleepers.

Assuming potential errors in the LCI data obtained for the production of the constituent sleeper materials, the sensitivity of the weighted findings to variations in this data can be ascertained. Based on an error range of ±40 per cent for the energy, water and emissions data used, Table 5.24 shows the possible effect of this data on the weighted environmental impacts associated with the sleeper production.

This sensitivity analysis shows that the potential errors associated with the data used for the study may have a significant impact on the findings, with a sensitivity of 40 per cent. In the worst-case scenario, whilst this may have no effect on the preferred sleeper type based on the energy use and global warming impact categories, it could potentially reverse the finding that reinforced concrete sleepers provide reduced water consumption compared to timber sleepers. Also, had the durability and anticipated service-life of the sleepers not been considered in the definition of the functional unit, then the findings may not have indicated the true extent to which reinforced concrete sleepers provide reduced energy and water consumption and lower greenhouse gas emissions compared to timber sleepers.

Table 5.24 Sensitivity analysis on weighting result for embodied energy, water and emissions data uncertainty, by railway sleeper type

Impact category	Weighting result	
	Reinforced concrete	Timber
Energy use (GJ)	2,601	7,514
Water use (kL)	3,525	7,214
Global warming (t CO_2-e)	328	1,740
Deviation (%)	±40	±40
Energy use range (GJ)	1,561 – 3,641	4,508 – 10,520
Water use range (kL)	2,115 – 4,935	4,328 – 10,100
Global warming range (t CO_2-e)	197 – 459	1,044 – 2,436
Sensitivity (%)	40	40

5.2.2 Case study 5: Road construction

This case study demonstrates how LCA can be used to select between competing or alternative materials used in road construction. A streamlined LCA approach has been taken, quantifying only energy consumption, in order to highlight the process used for selecting an environmentally preferred type of road construction.

Scope: The study quantifies the energy inputs associated with the construction and maintenance of four alternative road construction types in order to compare the total energy requirements associated with each type of road construction. The attributes of these road types are detailed in Table 5.25.

>   *Impact category*:   Energy use
>   *Category indicator*:   Gigajoules of energy

Functional unit: The functional unit for the assessment was a 1 m length of road over a period of 40 years. The trafficable width of the road was assumed to be 7 m with 2 m-wide shoulders on either side.

System boundary: The energy inputs associated with the initial construction and on-going maintenance of each road type were assessed. All energy requirements associated with the actual road construction process and all supporting processes and services upstream of this were included, as per Figure 5.13. Energy requirements associated with the installation of sub-surface drainage were assumed to be the same for all road types and thus not included in the analysis. The demolition stage was not included as very rarely are existing roads demolished; rather they are often maintained indefinitely into the future.

Table 5.25 Road type characteristics for 1 m length of road

Road type	Base	Sub-base
Continuously reinforced concrete (CRC)	2.09 m^3 of 32 MPa concrete; 1.38 m^3 of 5 MPa concrete 133 kg steel reinforcement	
Full-depth asphalt (FDA)	2.97 m^3 of asphaltic concrete	1.65 m^3 of stabilized earth[*]
Deep-strength asphalt (DSA)	2.22 m^3 of asphaltic concrete	Compacted graded earth
Deep-strength asphalt on bound sub-base (DSAB)	1.65 m^3 of asphaltic concrete	2.2 m^3 of stabilized earth[*]

*For road types FDA and DSAB, the sub-base of stabilized earth is assumed to be modelled adequately by 5 MPa concrete. No embodied energy is attributed to compacted graded earth, other than the amount of direct energy implied in the input-output model for the road construction process.

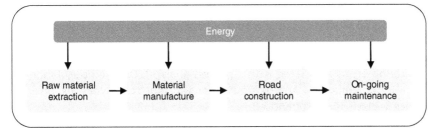

Figure 5.13 Life cycle stages and inputs considered for road construction life cycle assessment

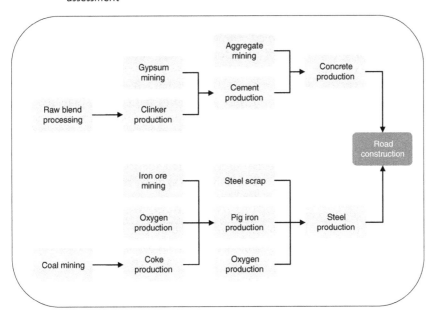

Figure 5.14 Processes included in process-based system boundary for continuously reinforced concrete (CRC) road construction life cycle assessment

Figure 5.14 shows the individual material production processes for which process data was obtained for the construction of the continuously reinforced concrete (CRC) road type.

5.2.2.1 Life cycle inventory of road construction and maintenance

The quantity of materials contained within a 1 m length of road (for all four road types) was determined based on an analysis of the individual road types (Table B.7, Appendix B). The energy requirements associated with the production of the road materials were calculated by multiplying the quantity of these materials by the hybrid material energy coefficients for the respective materials, which included some of the process data collected for the road material production processes. Additional process data was obtained from the latest SimaPro database, assuming Australian production. The resultant value represents all process-based energy requirements for the materials quantified, as well as those associated with the provision of goods and services upstream of and associated with these materials (avoiding the common upstream truncation errors associated with a process analysis). Detailed calculations of these initial embodied energy figures for the four road types are given in Tables B.7 to B.10 of Appendix B.

Input-output data was then used to fill any remaining data gaps (those sideways and downstream of the main materials and processes) in accordance with the input-output-based hybrid LCI approach outlined in Chapter 4. Using the energy-based input-output model developed by Lenzen and Lundie (2002), the pathways representing the individual materials quantified for the four road types were identified within the relevant economic sector (*Other construction*) and the sum of their *total energy requirements* (as per Table 5.26) subtracted from the *total energy requirement* of the sector. Other Stage One inputs into the *Other construction* sector that were considered not likely to be used in road construction were also subtracted (such as glass products and electrical equipment used in non-residential building construction, which is also part of the *Other construction* sector). The remaining figure represents any previously excluded energy inputs sideways or downstream of the main materials (including the direct energy required for road construction). This value in GJ per A$1,000 of sector output was then related to the specific functional unit by multiplying it by the total cost for a 1 m length of road to give the *remainder* value for the road length (Table 5.26). This *remainder* was then added to the initial embodied energy figure calculated in the previous step to determine the total energy requirements for the construction of each road type (as shown in Table 5.27 for the CRC road type).

Table 5.26 Calculation of remainder to fill sideways and downstream embodied energy data gaps for road construction

Road type	Sector	Sector TER (GJ/A$1000)[a]	Cost (A$)[b]	Sum TER of related paths (GJ/A$1000)[c] *	Remainder (GJ/m road) ((a-c) x b/1000)
CRC	Other construction	9.9798	596	5.074	2.924
FDA	"	"	858	4.695	4.534
DSA	"	"	630	3.904	3.828
DSAB	"	"	643	4.695	3.398

*This figure also includes inputs considered to not be used in road construction (e.g. glass products).

The annual energy consumption associated with maintenance of the roads was assumed to be equivalent to 4 per cent of the energy required for initial construction. Total embodied energy requirements of the four road types over a 40-year life cycle are shown in Table 5.28 and Figure 5.15.

Table 5.27 Calculation of initial embodied energy of CRC road construction

	Embodied energy (GJ/m road)
Process data for quantified road materials [a]	14.76
Input-output data used to fill *upstream* gaps for road materials [b]	12.58
Initial embodied energy [(a+b)]	27.34[c]
Input-output remainder (to fill *sideways* and *downstream* gaps) [(Table 5.26)]	2.924[d]
Total [(c+d)]	30.26[e]
Proportion of process data [(a/e)]	48.8%

Table 5.28 Life cycle energy requirements of four road types, per metre of road

Road type	Initial embodied energy (GJ)[a]	Annual maintenance energy (GJ)[b]	Life cycle energy (GJ)[(a+(b x 40))]
Continuously reinforced concrete (CRC)	30.26	1.21	78.68
Full-depth asphalt (FDA)	18.28	0.73	47.53
Deep-strength asphalt (DSA)	10.66	0.43	27.72
Deep-strength asphalt on bound sub-base (DSAB)	14.62	0.58	38.01

Note: Figures may not sum due to rounding.

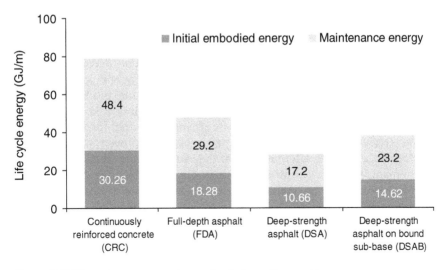

Figure 5.15 Life cycle energy requirements of alternative road types

5.2.2.2 Interpretation – evaluation of results

This case study shows that by assessing the environmental performance of a range of possible solutions, in this case for road construction, it is possible to identify a solution that may potentially result in the lowest environmental impact of all options considered (in this case the DSA road type). This does not mean however that this will necessarily be the preferred option. Simple and significant improvements to other options (such as the type and source of raw materials used or production process efficiency improvements) that may initially appear to have a greater environmental impact may in fact prove to be a better option. For example, the replacement of some or all of the cement used to manufacture the CRC road type with waste or lower impact cementitious material may significantly reduce the energy requirements for concrete production. For example, assuming a 50 per cent replacement of cement with flyash or a similar waste material could reduce the energy required for the roads' initial construction and on-going maintenance by as much as 21 per cent.

5.3 Renewable energy technology case studies

The rate of growth in the renewable energy sector has never been greater than at present. Renewable technologies are seen as a major part of the solution to reduce the environmental impacts, such as climate change, of fossil-fuel-based energy sources. There are many different technologies both well-developed and emerging that could be used to replace the various forms of energy that we currently consume. These range from using bio-waste and micro-algae to replace oil for use as transport fuel; wind turbines, solar photovoltaics and geothermal power plants to produce electricity needed to power our cities; and solar thermal systems to heat water and generate power. Considerable effort has gone into developing these technologies in an attempt to reduce the impacts caused by traditional technologies and fuel types. However, these new technologies bring with them additional issues of their own. For example, the potential for noise pollution and bird deaths from wind turbine operation (NWCC 2001), the depletion of raw materials and consumption of energy and water for their production, and the impact that some crop-based bio-fuel production has on food availability are all additional issues that must be considered.

Just as for any other product, LCA is an important tool for ensuring that these renewable energy technologies provide improved environmental performance over the alternatives that they are replacing. Impacts associated with manufacturing, construction, operation, maintenance and eventual disposal must also be considered, as for any element within the built environment, to ensure that the environmental benefits of producing renewable energy are not offset by the environmental burdens occurring throughout the life of these technologies. This section provides case studies dealing with two existing renewable energy technologies, namely a residential-scale photovoltaic system and a large multi-megawatt wind turbine.

5.3.1 Case study 6: Residential building integrated photovoltaic (PV) system

The purpose of this case study is to demonstrate the use of LCA for calculating the net energy production of solar photovoltaic (PV) systems and the potential greenhouse emissions avoided when compared to traditional coal-fired electricity production.

Scope: The study quantifies the energy inputs and outputs associated with the construction (including all supporting processes) and operation of a 1 kW residential-building-integrated PV system over a 25-year period. This system consists of 14 x 75 W crystalline silicon (c-Si) modules. The energy output of the PV modules was calculated based on the characteristics of the modules and climatic data for the location within which they were assumed to be located. The energy figures were then converted to global warming potential (in carbon dioxide equivalents (CO_2-e)) per GJ of equivalent coal-fired electricity to estimate the potential greenhouse gas emissions avoided with the use of the system.

Impact category:	Global warming
Category indicator:	Global warming potential (CO_2-e)

Functional unit: The functional unit for the assessment was a 1 kW PV system operating for 25 years. The embodied energy and total electrical output of the system were calculated and the net energy output determined. The net energy output was then converted to a quantity of greenhouse gas emissions (in GWP or CO_2-e) avoided per GJ of energy produced.

The details of the PV system are given in Table 5.29. The system consists of the grid-connected building-integrated crystalline silicon (c:Si) solar modules, necessary wiring and an inverter. The location of the PV system was assumed to be Sydney, Australia (33.9°S 151.2°E). The localized weather conditions, particularly the availability of solar radiation, determine the potential electrical output of the system.

Table 5.29 Photovoltaic system characteristics

	Photovoltaic system
Total rated power output	1,050 W (14 modules)
Module rated power output	75 W
Module dimensions	530 x 1,188 mm
Module weight	7.5 kg
Module average efficiency	11.9%
Inverter rated output	1,100 W
Insulated wiring length	20 m
Capacity factor	13.6%

System boundary: Only the energy inputs associated with the initial production (excluding on-going component replacement) and energy produced over the life of the PV system were assessed, including energy required for internal controls and system losses but excluding energy required for end-of-life refurbishment, disposal and possible reuse and recycling of components. All energy requirements associated with the actual production process and all supporting processes and services upstream of this were included, as per Figure 5.16 (such as, but not limited to, transportation and manufacturing of materials, and provision of capital equipment).

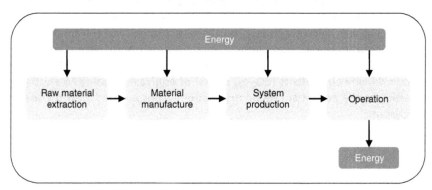

Figure 5.16 Life cycle stages considered for photovoltaic system life cycle assessment

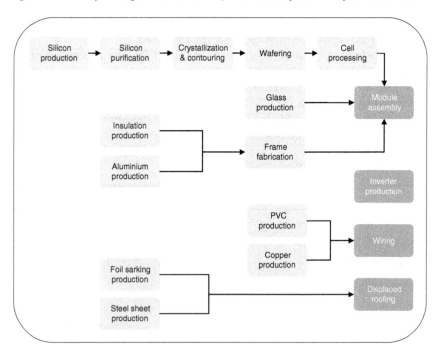

Figure 5.17 Processes included in process-based system boundary for photovoltaic system life cycle assessment

The building materials displaced as a result of integrating the modules into a residential roof were determined and included as an energy and emissions credit. The infrastructure associated with grid connection, beyond the electricity meter, was excluded. Figure 5.17 shows the individual production processes for which process data was obtained.

5.3.1.1 Life cycle inventory of photovoltaic system production and operation

PHOTOVOLTAIC SYSTEM PRODUCTION

The material quantities contained within the PV system were determined based on an analysis of the individual system components (Table B.11, Appendix B). The energy embodied in the modules, wiring and inverter was calculated by multiplying the material quantities for these components by the hybrid material energy coefficients for the relevant materials, which included some of the process data collected for the material production processes. Additional process data was obtained from a previous study of similar PV modules by Alsema *et al.* (1998). The resultant value represents all process-based energy requirements for the materials quantified, as well as all energy requirements associated with the provision of goods and services upstream of and associated with these materials (avoiding the common upstream truncation errors associated with a process analysis). Detailed calculations of this initial embodied energy figure for the PV system are given in Table B.11 of Appendix B.

Input-output data was then used to fill any remaining data gaps (those sideways and downstream of the main materials and processes) in accordance with the input-output-based hybrid LCI approach outlined in Chapter 4. From the energy-based input-output model developed by Lenzen and Lundie (2002), the pathways representing the individual materials quantified for the PV system (as per Figure 5.17) were identified within the relevant economic sector (*Other electrical equipment* for the PV system) and the sum of their *total energy requirements* (as per Table B.12, Appendix B) subtracted from the *total energy requirement* of the sector (17.689 GJ/A$1,000). The remaining figure represents any previously excluded energy inputs sideways or downstream of the main materials. This value in GJ per A$1,000 of sector output was then related to the PV system by multiplying it by the total cost of the system to give the *remainder* value for the PV system (Table 5.30), which was then added to the initial embodied energy figure calculated in the previous step to determine the total embodied energy of the PV system (Table 5.31). The energy embodied in the displaced roofing materials was calculated in the same manner (based on the input-output model for the *Residential building* sector) and subtracted from the total embodied energy of the PV system.

Table 5.30 Calculation of remainder to fill sideways and downstream embodied energy data gaps for photovoltaic system

Sector	Sector TER (GJ/A$1000)[a]	Element	Cost (A$)[b]	Sum TER of related paths (GJ/A$1000)[c]	Remainder (GJ/system) ((a−c) x b/1000)
Other electrical equipment		Modules (14)	11,970		
		Inverter	2,097		
		Wiring	50		
	17.689	Total system	14,117	5.3889	173.64
Residential building	10.633	Displaced roofing	1,050	0.1198	(11.04)
				Total	162.60

[c] From Table B.12.

Table 5.31 Calculation of initial embodied energy of photovoltaic system

	Embodied energy (GJ/system)
Process data for quantified PV system materials [a]	137.51
Input-output data used to fill *upstream* gaps for PV system materials [b]	27.38
Process data for quantified displaced roofing materials [c]	1.95
Input-output data used to fill *upstream* gaps for displaced materials [d]	6.28
Initial embodied energy [(a+b) − (c+d)]	156.67[e]
Input-output remainder (to fill *sideways* and *downstream* gaps) [(Table 5.30)]	162.60[f]
Total [(e+f)]	319.27[g]
Proportion of process data [(a/g)]	43.1%

PHOTOVOLTAIC SYSTEM NET LIFE CYCLE ELECTRICAL OUTPUT

The annual gross electrical output of the PV system was calculated, based on the characteristics and efficiencies of the PV modules and inverter as well as the climatic data for Sydney, Australia. Energy required for internal controls and system losses was assumed to be 1 per cent of the gross electrical output. The net energy output was determined by subtracting this energy requirement from the gross output. The net quantity of energy produced over the 25-year period is represented by Equation 5.1 and shown in Table 5.32.

Table 5.32 Net life cycle energy output of photovoltaic system

Annual gross electrical output (MJ)[a]	Annual system losses (MJ)[b]	Life cycle (years)[c]	Net life cycle energy output (MJ/25yrs)[((a−b) x c)]
4,536	46.8	25	112,230

$$LCE_{output} = \left(E_{output} - E_{operation}\right) \times Y \qquad \text{Equation 5.1}$$

Where LCE_{output} = net life cycle energy output; E_{output} = annual gross electrical output; $E_{operation}$ = annual energy required for internal controls and losses; Y = expected service-life of PV system in years.

5.3.1.2 Impact assessment – net life cycle greenhouse gas emissions avoided

The calculated embodied energy of the PV system was converted to GWP impacts using the results from the LCI and the characterization factor of 60 kg CO_2-e per GJ. To provide an indication of the environmental benefit due to the produced energy output, the potential emissions avoided were calculated based on the life cycle output of the PV system, to represent the greenhouse gas emissions associated with the production of a similar quantity of energy from a coal-fired power station. The net life cycle energy output (112,230 MJ) was converted to primary energy terms using a factor of 3.1 to reflect the total energy required to produce electricity from black coal in Australia. Emissions factors for black coal-fired electricity were then used to convert the energy output into emissions terms. GWP factors for the three main greenhouse gases (CO_2, CH_4 and N_2O) were then used to convert the emission figure into CO_2-e terms. The net life cycle emissions avoided with the use of the PV system were then able to be determined, as in Table 5.33.

Table 5.33 Net life cycle greenhouse gas emissions of photovoltaic system manufacture and operation

Life cycle stage	Quantity of primary energy (GJ)	GHG	Emissions factor (kg per GJ)[^]	GWP*	Total emissions produced/(avoided) (t CO_2-e)
Embodied energy	319.27	CO_2-e	60		19.16
Life cycle energy output (25 years)	347.9	CO_2	88.2	1	(30.68)
	"	CH_4	0.03	25	(0.26)
	"	N_2O	0.2	298	(20.73)
			Total net emissions avoided		32.51
Net emissions avoided per GJ produced electricity					0.29

[^]*Source:* Department of Climate Change 2008: 202.
*Global Warming Potential, *Source:* IPCC 2007b: 212.

The potential avoided greenhouse gas emissions associated with the use of the 1 kW PV system in Sydney, Australia is 32.51 t CO_2-e over a 25-year period, which equates to 0.29 t CO_2-e per GJ of electricity produced. The potential GWP benefits will obviously differ from country to country depending on the conventional fuels and processes used for energy production in those countries. The emissions associated with the construction of the coal-fired power station have also been excluded, which may provide an underestimation of the potential avoided emissions associated with the use of a PV system, particularly if being compared to the construction of a new coal-fired power station.

5.3.1.3 Interpretation – evaluation of results

The LCA of the PV system showed that it may take over 20 years for the energy requirements associated with the manufacture of the system to be repaid by the energy being produced by it. Analyzing the results, it is clear that the greatest impacts associated with the PV system relate to the production of the silicon used in the PV modules (see Table B.11, Appendix B). This type of analysis is useful as it can be used to direct any environmental improvement efforts towards the processes found to be most significant. In this case, this may involve improving resource efficiency of the manufacturing process, sourcing alternative materials for solar cell production or changing or improving other aspects of the manufacturing processes used.

Assuming potential errors in the LCI data obtained for the production of the PV system, the sensitivity of the findings to variations in this data can be ascertained. Based on an error range of ±20 and ±50 per cent for the process and input-output components of the embodied energy data used, Table 5.34 shows the possible effect of the data used on the avoided greenhouse gas emissions associated with the PV system.

Table 5.34 Sensitivity analysis for embodied energy data uncertainty on avoided emissions for photovoltaic system

	Process data	Input-output data	Total emissions produced/(avoided) (t CO_2-e)
Original embodied energy GWP, by data type (t CO_2-e)	8.13	11.02	19.16
Uncertainty (%)	±20	±50	
Embodied energy GWP range (t CO_2-e)	6.5 – 9.8	5.5 – 16.5	$12^a – 26.3^b$
Life cycle avoided emissions from energy output			51.67^c
Total net emissions avoided			$25.37^{(c-b)} – 39.67^{(c-a)}$
Original total net emissions avoided			32.51
Sensitivity (%)			22

The sensitivity analysis shows that the potential errors associated with the data used for the study may have a significant impact on the findings, with a sensitivity of 22 per cent. Whilst this may appear to be a considerable point of concern, even based on the lowest possible value for the net emissions of the PV system, it should still result in the avoidance of around 25 tonnes of emissions over a 25-year period.

Of course, as with any LCA study, these findings may also vary depending on certain assumptions made and the data used for the analysis. These may include variations to the manufacturing technologies used, location of production and installation, system efficiencies, operating conditions, chosen system boundaries and the quality and completeness of energy data obtained.

5.3.2 Case study 7: Wind turbine

This case study assesses the energy-related global warming impacts resulting from the construction and operation of a single on-shore wind turbine in order to demonstrate how strategies for environmental improvement and the potential net greenhouse gas emissions benefit of the turbine can be identified.

Scope: The study quantifies the energy inputs associated with the construction, component replacement (including all supporting processes) and operation of the turbine over a 20-year period. Energy use has been converted to primary energy terms to account for the impacts associated with its production. The energy output of the turbine during this time has also been calculated based on the characteristics of the turbine itself and climatic data for the location within which it was assumed to be located. These energy figures were then converted to global warming potential (in carbon dioxide equivalents (CO_2-e)) per GJ of equivalent coal-fired electricity to estimate the potential greenhouse emissions avoided with the use of the turbine.

Impact category: Global warming
Category indicator: Global warming potential (CO_2-e)

Functional unit: A single 3.0 MW horizontal axis, three-blade wind turbine was assessed over a lifetime of 20 years. The details for the wind turbine are shown in Table 5.35. The turbine consists of the base, tower, rotor (hub and blades) and nacelle (containing the gearbox, generator, brakes and other electrical components) (Figure 5.18) and is located on the south-west coast of Victoria, Australia.

System boundary: Only the energy inputs associated with the construction (initial and on-going component replacement) and operation stages of the wind turbine were assessed, excluding energy required for end-of-life refurbishment, demolition and possible reuse and recycling of components. All energy requirements associated with the actual construction process and all supporting processes and services upstream of this were included as per

Figure 5.19, as well as the energy required for operating the turbines (for internal controls and day-to-day maintenance). The access roads, transformers and components associated with grid connection were excluded.

Table 5.35 Wind turbine characteristics

	Wind turbine
Total rated power output	3,000 kW
Capacity factor	33%
Cut-in wind speed	4 m/s
Cut-out wind speed	25 m/s
Hub height	80 m
Blade length / rotor diameter	44 m / 90 m

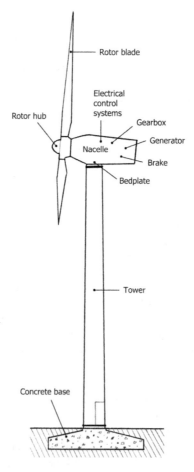

Figure 5.18 Wind turbine components

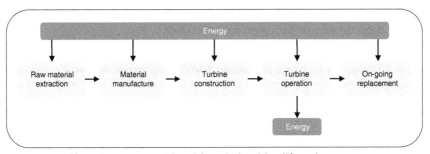

Figure 5.19 Life cycle stages considered for wind turbine life cycle assessment

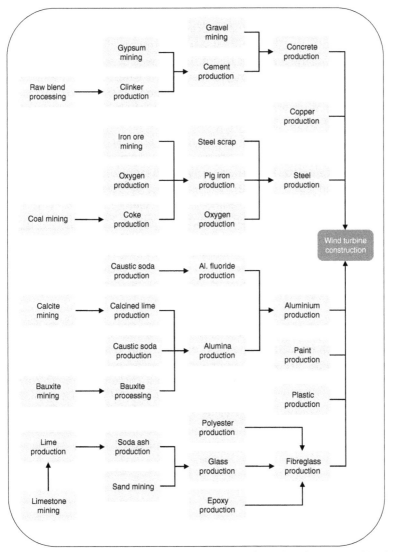

Figure 5.20 Processes included in process-based system boundary for wind turbine life
cycle assessment

Figure 5.20 shows the individual production processes for which process data for the wind turbine construction was obtained.

5.3.2.1 Life cycle inventory of wind turbine construction and operation

WIND TURBINE CONSTRUCTION

The material quantities contained within the turbine were determined based on an analysis of the individual turbine components (Table B.13, Appendix B). The energy embodied in the wind turbine was calculated by multiplying these material quantities by the hybrid material energy coefficients for the relevant materials, which included the process data collected for the material production processes. The resultant value represents all process-based energy requirements for the materials quantified, as well as all energy requirements associated with the provision of goods and services upstream of these materials (using input-output data to avoid the common upstream truncation errors associated with process analysis). Detailed calculations of this initial embodied energy figure for the turbine are given in Table B.13 of Appendix B.

As per the input-output-based hybrid LCI approach, any remaining data gaps (those sideways and downstream of the main materials) were then filled with input-output data. From the energy-based input-output model developed by Lenzen and Lundie (2002), the pathways representing the individual materials quantified for the wind turbine were identified (Table B.14, Appendix B) and their *total energy requirements* subtracted from the *total energy requirement* of the relevant economic sector. As the construction of the turbine involved a number of components coming from different industries (for example, the base from the *Other construction* sector and the tower from the *Structural metal products* sector), the sum of the *total energy requirements* of the pathways representing the materials for each of these components was subtracted from the *total energy requirement* of the appropriate sector (as listed in the first column of Table B.14, Appendix B). The remaining pathways (those not subtracted from the *total energy requirement*) for each of these sectors completes the system boundary for any previously excluded inputs sideways or downstream of the main materials. These values were converted to energy terms by multiplying them by the cost of each component and summed to give the *remainder* value for the wind turbine (Table 5.36). This *remainder* was then added to the initial embodied energy figure calculated in the previous step to determine the total embodied energy of the turbine (Table 5.37).

Table 5.36 Calculation of remainder to fill sideways and downstream embodied energy data gaps for wind turbine

Sector	Sector TER (GJ/A$1000)[a]	Element	Cost (A$'000)[b]	Sum TER of related paths (GJ/A$1000)[c]	Remainder (GJ/turbine) ((a−c) x b/1000)
Other construction	9.98	Base	396.9	1.653	3,305
Structural metal products	18.16	Tower	705.6	7.372	7,612
Other machinery and equipment	15.06	Nacelle	2,730	6.324	23,858
Plastic products	21.66	Blades	1,091	10.750	11,915
Fabricated metal products	17.66	Hub	326	5.658	3,917
				Total	50,607

[c] From Table B.14.

Note: Figures may not sum due to rounding.

Table 5.37 Calculation of initial embodied energy of wind turbine

	Embodied energy (GJ/turbine)
Process data for quantified wind turbine materials[a]	18,716
Input-output data used to fill *upstream* gaps for wind turbine materials[b]	15,041
Initial embodied energy [(a+b)]	33,757[c]
Input-output remainder (to fill *sideways* and *downstream* gaps) [(Table 5.36)]	50,607[d]
Total [(c+d)]	84,364[e]
Proportion of process data [(a/e)]	22%

WIND TURBINE COMPONENT REPLACEMENT

The energy associated with the replacement of parts during the expected 20-year lifetime of the turbine was assumed to be equivalent to the replacement of half of the gearbox over the service-life (based on Vestas Wind Systems 2006). Gear oil and grease were also assumed to be replaced at five-year intervals (at year 5, 10 and 15). The initial embodied energy for these components was calculated based on 50 per cent of the embodied energy for the gearbox and three times the energy embodied in the oil. The *remainder* was then calculated to account for the sideways and downstream data gaps based on half of the price of the gearbox and the *total energy requirement* for the *Other machinery and equipment* sector. These figures were then summed to give the total recurrent embodied energy requirement for the wind turbine (Table 5.38).

Table 5.38 Calculation of recurrent embodied energy of wind turbine

	Embodied energy (GJ/turbine)
Initial energy embodied in gearbox[a]	2,276
Recurrent embodied energy of parts replacement (50% of [a])[b]	1,138
Replacement of oil and grease (three no.)[c]	69.3
Recurrent embodied energy [(b+c)]	1,207.3[d]
Input-output data for replacement parts:	
Total energy requirement*	6,975[e]
Inputs covering process data	240.3[f]
Input-output remainder (to fill *sideways* and *downstream* gaps) [(e–f)]	6,735[g]
Total [(d+g)]	7,942

*Based on half of the price of the gearbox (A$463,050) and the TER of the *Other machinery and equipment* sector (15.06 GJ/A$1,000).

WIND TURBINE NET LIFE CYCLE ELECTRICAL OUTPUT

The annual gross electrical output of the wind turbine was calculated, based on the characteristic power curve and hourly wind data for the chosen location. A capacity factor of 33 per cent was assumed, accounting for the efficiency of the turbine as a proportion of its rated output. Energy required for internal controls, day-to-day maintenance and system losses was assumed to be 10 per cent of the gross electrical output. The net energy output was determined by subtracting this energy requirement from the gross output. The net quantity of energy produced over the 20-year period is represented by Equation 5.2 and shown in Table 5.39.

$$LCE_{output} = \left(E_{output} - E_{operation}\right) \times Y$$

Equation 5.2

Where LCE_{output} = net life cycle energy output; E_{output} = annual gross electrical output; $E_{operation}$ = annual energy required for internal controls, maintenance and losses; Y = expected service-life of turbine in years.

Table 5.39 Net life cycle energy output of wind turbine

Annual gross electrical output (GJ)[a]	Annual system losses (GJ)[b]	Life cycle (years)[c]	Net life cycle energy output (GJ/20yrs)[((a–b) x c)]
118,494	11,849	20	2,132,900

5.3.2.2 Impact assessment – net life cycle greenhouse gas emissions avoided

It is important that renewable energy systems provide a net environmental benefit compared to conventional energy production systems (such as coal, nuclear or gas-fired power plants). Net GWP benefits can be determined by subtracting the GWP impacts (associated with embodied energy and operation) from the GWP benefits (attributable to the emissions-free electricity output of the turbine). The embodied energy requirements associated with the construction and on-going component replacement of the wind turbine were converted to GWP impacts using the results from the LCI and the characterization factor of 60 kg CO_2-e per GJ of energy. To provide an indication of the environmental benefit due to the produced electrical output, the potential emissions avoided were calculated based on the life cycle output of the wind turbine, to represent the greenhouse emissions associated with the production of a similar quantity of energy from a coal-fired power station. The net life cycle energy output figure (2,132,900 GJ) was converted to primary energy terms using a factor of 3.4 to reflect the total energy required to produce electricity from brown coal in Victoria, Australia. Emissions factors for brown coal-fired electricity were then used to convert the electrical output into emissions terms. GWP factors for the three main greenhouse gases (CO_2, CH_4 and N_2O) were then used to convert the emission figure into CO_2-e terms. The net life cycle emissions avoided with the use of the wind turbine were then able to be determined, as in Table 5.40.

The potential avoided greenhouse gas emissions associated with the use of the wind turbine were found to be 1,532,944 t CO_2-e over a 20-year period, which equates to 0.719 t CO_2-e per GJ of electricity produced. The potential GWP benefits will obviously differ from country to country depending on the

Table 5.40 Net life cycle greenhouse gas emissions of wind turbine construction and operation

Life cycle stage	Quantity of primary energy (GJ)	GHG	Emissions factor (kg per GJ)^	GWP*	Total emissions produced/(avoided) (t CO_2-e)
Embodied energy	92,306	CO_2-e	60		5,538
Life cycle energy output (20 years)	7,251,860	CO_2	92.7	1	(672,247)
"		CH_4	0.01	25	(1,813)
"		N_2O	0.4	298	(864,422)
		Total net emissions avoided			1,532,944
	Net emissions avoided per GJ produced electricity				0.719

^Source: Department of Climate Change 2008: 202.
*Global Warming Potential, Source: IPCC 2007b: 212.

conventional fuels and processes used for energy production in those countries. The emissions associated with the construction of the coal-fired power station have also been excluded, which may provide an underestimation of the potential avoided emissions associated with the use of a wind turbine, particularly if being compared to the construction of a new coal-fired power station. Of course, as with any LCA study, these findings may also vary depending on certain assumptions made and the data used for the analysis. These may include variations to the manufacturing technologies used, location of production and installation, system efficiencies, operating conditions, chosen system boundaries and the quality and completeness of energy data obtained.

5.3.2.3 Interpretation – evaluation of results

This case study shows that a single 3.0 MW wind turbine can make a significant contribution towards reducing the greenhouse gas emissions associated with electricity production. Assuming that the electricity requirements associated with operating a typical house result in the production of 12 tonnes of greenhouse gas emissions on an annual basis, this single wind turbine is capable of offsetting the emissions associated with over 6,300 households. It is also important to note that whilst the initial energy requirements associated with the construction of the turbine may be considered to result in a negative impact on the environment (especially if the manufacturing processes use fossil-fuel-based energy), they are recovered by the produced electrical output of the turbine in slightly less than a year. Although reductions in embodied energy may provide additional benefit, this shows that excessive time spent on this area may not provide significant benefits, and must be balanced against a range of other factors such as the cost, durability and maintenance of the turbines.

Assuming potential variations in the characterization factor used to convert the embodied energy consumption to GWP terms, the sensitivity of the findings to variations in this figure can be ascertained. Based on a possible variation of +33.3 per cent for this characterization factor, Table 5.41 shows the possible effect of this variability on the net life cycle avoided greenhouse gas emissions associated with the wind turbine.

Table 5.41 Sensitivity analysis on GWP characterization factor for calculating embodied energy-related emissions of wind turbine

	Original	Variation	Difference
Characterization factor (kg CO_2-e/GJ)	60	80	+20
Embodied energy GWP (t CO_2-e) [a]	5,538	7,384	+1,846
Life cycle avoided emissions from energy output [b]	1,538,482	1,538,482	
Total net emissions avoided [(b-a)]	1,532,944	1,531,098	−1,846
Sensitivity (%)			0.12

The sensitivity analysis shows that variations to the characterization factor used for converting the embodied energy of the wind turbine to GWP may have very little influence on the overall outcomes of the analysis, with a sensitivity of only 0.12 per cent.

The previous two case studies on renewable energy technologies have demonstrated how LCA can be used to assess whether these technologies will actually provide a net environmental benefit, including the consideration of the resources required for their manufacture and on-going maintenance. Whilst these case studies have been streamlined in this instance to focus on only energy-related impacts (which is warranted, as the primary purpose of these systems is to reduce the greenhouse gas emissions associated with energy production and use), the benefits that these technologies provide in reducing fossil fuel depletion should also not be overlooked. Other impacts associated with the raw materials and manufacturing processes used for these technologies would also be considered when conducting a more comprehensive LCA study, covering a broader range of environmental parameters (including pollutant releases and natural resource depletion).

5.4 Summary

This chapter has provided a number of case studies across some of the various elements of the built environment to demonstrate the application of LCA in the built environment context. The findings from these studies provide useful information to help decision makers make more informed environmental choices to improve the environmental performance of the built environment.

The impact that some of the limitations of the LCA approach can have on assessing environmental performance has been demonstrated. The final chapter places the information that has been obtained from these and other LCA studies in a broader environmental improvement context. It addresses what can and should be done by the various stakeholders involved in the built environment to minimize environmental impacts and shows how LCA is just one tool that can be used to assist in achieving this goal.

6 Opportunities for reducing the environmental impact of the built environment

Ultimately, our goal must be to avoid all negative impacts on the natural environment that result from our activities, including those resulting from the design, construction and operation of the built environment. Only then can we hope to avoid further environmental pollution and depletion of non-renewable resources and reverse the environmental degradation that has already occurred. LCA is an essential tool for identifying the impacts that are occurring or may occur and for informing better decisions that go as close as possible to avoiding any negative impacts on the environment. The previous chapter demonstrated the type of information that an LCA can provide to better inform environmental decision making. LCA can go a long way towards supporting decision-making processes but it does not suggest the preferred way forward. Professional judgement and knowledge is still an essential element for making the best environmental decisions in particular geographic and operational contexts.

Knowledge gained from previous LCA studies helps to light the pathway towards improved environmental performance in the built environment. There is a range of simple and often easily employed strategies that will almost always lead to a reduction in the environmental impacts of the built environment. This chapter briefly discusses what some of the key stakeholders in the built environment – such as designers, owners and occupants – can and should be doing to minimize these impacts. Where the environmental benefits are not as obvious, it is essential that LCA is used to better understand the long-term implications of choosing particular alternative materials or components or changing to alternative practices or processes related to resource extraction, material processing and construction.

Governments also have a key role to play and can use a range of policy and educational strategies to lead environmental improvement efforts. However, care must be exercised as some of these may have adverse economic and social impacts and therefore must be carefully balanced with broader societal desires to maintain or improve current standards of living. The implications for industry and broader society of some of the possible strategies that may be employed by governments to reduce environmental impacts are also presented in this chapter. Finally, opportunities and needs for further research are discussed, particularly in terms of the future development of LCA and its integration into industry.

6.1 Reducing environmental impacts of the built environment

In order to reduce or avoid impacts on the natural environment associated with the construction, use and eventual demolition and disposal of buildings and other elements within the built environment, it is first necessary to quantify what these impacts are, where they occur and how significant they might be.

Current knowledge of many of the potential impacts and how they can be managed is growing. There is a range of simple and often cost-effective strategies that are increasingly being employed to minimize the impact of both existing and new built environment assets. For existing buildings these often relate to maintaining the structural and servicing elements of buildings and other infrastructure to minimize the need for substantial component replacement (and thus additional resource, energy and water consumption and waste production). Other strategies include improving the operational efficiency of buildings and advocating behavioural change to reduce operational environmental impacts from energy and water consumption and waste production of building occupants. Even greater benefits are possible for new buildings by designing them to maximize operational energy and water efficiencies, by using durable materials to maximize their service-life and by using renewable or recycled materials to reduce the impacts associated with demand for non-renewable virgin materials. However, despite these current efforts, this knowledge does not yet provide a thorough understanding of the myriad of different environmental impacts occurring across the life cycle of a built environment asset. This is particularly the case for the indirect impacts, mainly due to the difficulties associated with their identification and subsequent quantification. Only with this thorough understanding can the environmental performance of the built environment be optimized in a manner that results in a truly sustainable outcome. How far we are along the pathway towards zero long-term human impacts on the environment can then be better understood.

LCA is one of the best tools available for achieving this and for providing a more complete view of the environmental impacts that may result from particular resource extraction, manufacturing, construction and demolition processes and practices. LCA can be useful to a certain point as it provides information that supports decision-making processes. However, it does not provide definitive answers and care should always be taken when interpreting and using the findings of any LCA study as even minor variations in the assumptions, data or weightings used can change the study outcomes, as has been demonstrated earlier in Chapter 5. An LCA can provide information on the significance and type of impacts, which can then be used to compare alternatives and make decisions where there are competing choices or target and prioritize environmental improvement or management strategies to particular processes or life cycle stages. An LCA should also be part of any decision-making process due to the range of possible design options that may provide solutions to environmental issues. The direct and indirect impacts of any decision can then be balanced in order to choose the best option from a life cycle environmental perspective.

Despite the significant benefits that an LCA study offers, expert or professional judgement and understanding of the product system under analysis is still essential for interpreting the findings and implementing appropriate environmental management or improvement strategies. Essentially, this means that it is up to those involved in the design and management of built environment assets (designers, developers, facility managers and owners) to make the correct decisions, based on the best available information and their previous experience.

6.1.1 Barriers to optimizing environmental outcomes

Existing research has shown that those involved in the design and construction of the built environment (designers, builders, engineers and consultants) consider environmental improvement important amongst the issues facing the industry (Jackson *et al.* 2010). However, numerous barriers still exist that often work against those willing to lead the necessary shift from traditional to more sustainable approaches to the construction and operation of the built environment. Developers of built assets are often reluctant to shift away from the tried and tested approaches for procuring their projects. This is often due to the perceived risks that they envisage might ensue from taking a more environmentally conscious approach. These can include the risks of reduced financial return due to a lack of information on the potential life cycle costs and environmental benefits over the long term, as well as hidden costs and split incentives (Sorrell *et al.* 2004).

6.1.1.1 Market demands and split incentives

Non-owner developers, designers and even tenants are often downstream in the supply chain that is the driver for new building and infrastructure projects. Each of these stakeholders can be constrained in their efforts to improve the environmental performance of the built environment by what the market demands. Ultimately decisions are made by those that hold the greatest financial stake in a project: namely the owner or investor. Their decisions will be driven by their own needs and desires or those of the market to which they will be selling the project. Market demand is slowly shifting (more so in some countries than in others) to environmentally preferred alternatives for housing and other building types. However, markets in the developed world particularly still tend to act on their perceived right or ability to build larger houses and buildings over any desires they may have to reduce their environmental impact.

Also, developers that dispose of a project shortly after construction have little impetus for spending more upfront when they are unlikely to benefit from long-term financial savings. For example, while the developer may pay the capital cost of incorporating many operational efficiency measures, this may not result in increased profits, as many of the operational savings will accrue to owners or tenants by way of reduced operating or occupancy costs (WBCSD 2007; Myers *et al.* 2008). This split incentive may change as the market develops a greater appreciation for the many benefits that more resource efficient buildings and infrastructure systems can offer.

6.1.1.2 Accessibility to consistent environmental data

As is the case with the application of LCA, a lack of accessible, consistent environmental data can be a major barrier to implementing environmental best practice in built environment projects. Where environmental performance information is available, it is often sourced from disparate sources. This makes comparisons of alternative solutions unreliable and risky as this data is often compiled using different assessment approaches, system boundaries and can cover a variety of environmental parameters and life cycle stages. Despite the best intentions, the most intelligent resolutions and designs will not be reached if they are based on unreliable or incomplete data.

6.1.1.3 Lack of regulation

Current government regulation in many countries focuses on setting minimum environmental performance standards rather than encouraging maximum environmental performance. For example, Australia has the *5 Star standard* (Building Commission 2008) which requires all new housing to be designed within maximum heating and cooling energy loads per square metre of floor area. There is no penalty for building larger houses or incentive for building smaller houses that would also help to limit the resources embodied in those houses. In many cases like this, as there is often no regulatory-driven incentive to maximize the environmental performance of buildings beyond this minimum threshold many designers and owners will aim to only just comply with maximum energy load requirements rather than strive to optimize the building design to achieve the lowest energy loads possible.

Other than attempting to address some of the above mentioned real and perceived barriers that currently act as impediments to creating a more environmentally conscious built environment, there are also a number of other things that the numerous stakeholders in the built environment can do to support this cause.

6.1.2 What can designers do?

Designers of built environment assets have a professional responsibility to meet their clients' needs but also provide them with reliable and timely information. They must also respond to pressing social and environmental issues which are often beyond traditional client expectations and desires. With our growing understanding of how the built environment impacts the natural environment, this must also extend to leading and implementing environmental best practice in everything they do. These professional responsibilities increasingly must also extend to educating clients so that the desire for environmental improvement is intrinsic to the way in which broader society also operates.

'We need a new environmental consciousness on a global basis. To do this, we need to educate people.'

Mikhail Gorbachev (n.d.)

6.1.2.1 Education and training

To facilitate this leadership role, designers must have more than just a basic understanding of the principles and issues associated with environmental design and improvement. They must continually strive to educate themselves so that they can provide the best possible service and advice to their clients and maximize the environmental performance of their own processes as well as their clients' projects. Designers' actions are of particular importance as their decisions can have long-term consequences on the environment. Most of the impacts occurring across the life cycle of the built environment are locked in during the design stage and due to the extended service-life of built environment assets they can persist for many decades.

There are numerous avenues through which designers can reduce the environmental impacts of built environment projects. However, many of these cannot be done without closely working with clients, who will have their own needs, desires and expectations for the project. Some clients will be happy to leave most of the decisions up to the designer; however, just as many will want a say in the materials that are used, the size of the project and some of the operational characteristics (such as a desire for tightly controlled internal building temperatures that can usually only be met by energy-intensive HVAC systems). Unless the client and designer are both engaged equally in optimizing the environmental performance of the project, then the project will ultimately fail to achieve the best environmental outcomes. Currently this tends to happen frequently as design solutions that contribute to improving environmental performance are often the first to be excluded when budget constraints begin to appear.

One of the biggest challenges can be convincing a client to take a more environmentally conscious approach to a project. This can be particularly difficult with a client that is unaware of the many long-term benefits of doing so or is initially unwilling due to misconceptions, a risk-adverse attitude or a lack of understanding of the benefits that it will eventually provide. Usually a project will end up as a compromise between a designer's ambitious goals and a client's willingness to take this approach, but this process can be made much easier when a client is fully engaged.

6.1.2.2 Maximize resource efficiency

Other than educating and working with clients to achieve the best possible environmental outcomes, the focus of environmental improvement for designers must be on minimizing non-renewable resource consumption and maximizing the value of those resources that are (in new built assets) and have already been (in existing built assets) used. Designers of built environment assets have a key and highly influential role to play in achieving this.

Designs for new buildings and other built assets should be rationalized as much as possible to avoid excessive or unnecessary resource consumption. This involves reducing building size as much as practicable, improving planning and circulation efficiencies and eliminating those details and elements that are not considered absolutely essential for functional, safety, regulatory or design reasons. This can sometimes pose a difficult dilemma for designers as it works against traditional market demands for large buildings and the typical basis on which they earn their income. Designers should strive to specify materials that have the lowest life cycle environmental impacts. This is where LCA is of greatest benefit and an essential component of improving the environmental performance of built assets. A thorough understanding of the full impacts associated with the use of a particular material is necessary in order to ensure the best environmental outcomes are being achieved.

Design for durability is important; however, there also needs to be a consideration of the likely time-in-service so that materials with greater embodied resources are not chosen to maximize durability well beyond the time for which they are going to be used. In this situation, a material with a lower embodied resource requirement might be just as suitable. For example, the specification of carpet for a retail tenancy should factor in that this carpet is most likely to be replaced not due to wear but more so due to changing trends or new tenants well before the anticipated life expectancy of a highly durable commercial grade carpet. This is one situation where designers must also seriously consider changing the way things are currently done to minimize wastage and maximize resource value.

6.1.2.3 Understand the broad range of potential impacts

Whilst the specification of materials that are reused or recycled can usually help to reduce impacts compared to the alternative solution of using virgin materials, designers should understand that the use of these materials will also result in a range of environmental impacts. Reusing materials can reduce the impacts associated with the depletion of finite resources; however they often require additional resources (e.g. energy and water) to prepare them for their new use or to reprocess them into new materials. For example, recycling glass requires around 75 per cent of the energy needed to produce glass from virgin materials as most of the energy is used in the melting of the glass. In contrast, to recycle aluminium only 5 per cent of the energy needed to produce aluminium from virgin raw material is required (The EarthWorks

Group 1990). There is also a chance that these materials may be less durable than new materials and thus need replacing on a more frequent basis. This can erode some of the potential benefits from the use of recovered materials.

It should now be clearly evident that the specification of materials that have minimal environmental impacts can be an extremely complex process. How can designers be expected to know what the best choices may be without significant support and the detailed information needed to make these decisions? It is no wonder that the best choices are not always made even despite the other competing factors (such as cost and client desires) that play a part in this process. In light of this, designers must strive to seek advice and assistance (through education, experts and environmental consultants) to better inform their design decisions and improve the environmental performance of their projects.

6.1.2.4 Optimize opportunities through design

Designers can not only have a significant influence on the impacts associated with a building's construction and operation but can also influence many of the longer-term impacts of a building. By designing buildings for adaptive reuse, where they can be modified in order to prolong their useful life beyond what might have otherwise been possible under their initial use, the life of buildings can be extended and the value of the resources used in the materials can be maximized.

Designers can also help to facilitate the reuse and/or recycling of materials when buildings need to be demolished or refurbished. By designing buildings for ease of disassembly, they can greatly encourage and maximize the opportunities for materials to be recovered at the end of a building's life rather than being sent to landfill. This could include consideration of the way in which connections between materials are made (such as screwing or bolting rather than gluing or nailing) or selection of construction methods that allow for the easy separation of materials.

Whilst designers have a major say in many of the impacts that a built asset will have over its life, through the materials and systems that are specified, the owners and occupants also have a considerable say in the extent to which some of these impacts occur. These relate mostly to the impacts associated with the energy and water consumed and waste produced during an asset's operation or use, which can be particularly significant for buildings.

6.1.3 What can owners and occupants do?

The owners and occupants of a built asset can take numerous forms (developer, investor, purchaser and tenant) and, depending on their association with an asset, they can have varying levels of influence over, and interest in, its life cycle environmental impacts. For example, whilst tenants can often influence the broader construction industry by demanding only buildings that meet certain minimum environmental standards, they are often restricted or have little incentive to make significant physical changes to buildings to improve their performance (such as installing fixed shading devices, insulation and more efficient fixed heating and cooling systems).

Also, developers and owners that intend to sell or lease out a built asset shortly after construction often lack an incentive to invest in improving their environmental performance. This is beginning to change now that many tenants are demanding minimum environmental standards in the buildings they lease. For example, the Australian Government has set minimum energy performance standards for new, refurbished or newly leased government-occupied buildings over 2,000 m^2 (Australian Greenhouse Office 2007: 19).

Owners that choose to occupy a building (owner-occupiers) or maintain ownership of an asset tend to have a greater incentive to implement strategies that will improve its environmental performance. This is particularly the case where they will personally benefit from direct and potentially long-term cost savings from reduced energy and water consumption or from investing extra in maintenance to prolong the life of the asset and reduce the frequency of any major or costly repairs. In order to maximize the opportunities for improving the environmental performance of built assets, owners should aim to:

- choose designers (for new assets or upgrades to existing assets) that have respected environmental standards, credentials, knowledge and intentions, that may be evidenced through previous work
- demand the best possible environmental outcomes when new assets are being built or existing ones upgraded
- take a life cycle view to financing, developing and/or owning built assets
- facilitate cultural change within an organization or behavioural change amongst occupants
- monitor performance and look to continually improve operations.

Maintaining assets to prolong their life can also be an important approach to minimizing environmental impacts. Whilst regular maintenance can be seen as a significant upfront cost impost, taking a life cycle view in this regard is also important. Regular maintenance more often than not will prolong the life of an asset, extending the time until replacement or major improvements are needed. This can help to maximize the value of resources and diminish the need for the replacement or major refurbishment of a building or infrastructure system and the additional impacts that come with this.

The responsibility for designing and operating buildings and other built assets in a more environmentally conscious manner does not lie solely with designers and occupants, however. The other stakeholders in the built environment also have a significant role to play. Other than the influence that industry and society can have, governments can also have a considerable impact on encouraging a more environmentally responsive approach to the design and operation of the built environment.

6.1.4 What can governments do?

Governments have certain powers that other stakeholders in the built environment do not. They have the ability to influence (or even control) market demands and expectations. With this power, governments can choose to implement policies to achieve improvements to existing environmental standards. Governments have a range of means by which they can encourage or enforce better environmental standards and achieve substantial reductions in environmental impacts. These include:

- regulating minimum environmental performance
- establishing resource (energy/water/waste) efficiency and disposal standards
- providing training and education to existing and future built environment professionals, consumers and industry
- funding research into and promoting the development of innovative materials, construction practices, environmental data and assessment tools
- providing incentives, including financial incentives (direct and indirect)
- providing penalties for poor environmental performance or environmental damage and to encourage improvement.

6.1.4.1 Incentives

Direct incentives can help to encourage or reward the uptake of environmentally preferred solutions, strategies and technologies by providing a cash rebate, subsidy or grant to offset any cost premiums that may otherwise restrict their broader uptake. There is also a range of factors that can indirectly encourage the uptake of environmentally preferred approaches to built asset design, construction and operation. For example, increasing energy and water prices, waste disposal charges and resource costs have already been shown to encourage greater resource conservation and better environmental practice.

6.1.4.2 Penalties as incentives

Penalizing poor environmental performance – such as inefficient resource consumption – through fees or taxes is another approach that can be used in an attempt to encourage environmental improvement. However this is only likely to work on a widespread basis where the penalties outweigh the risks and costs associated with changing current practice. This has the effect of indirectly creating an incentive to reduce consumption, especially where financial benefits are likely. This approach is most likely to lead to greater environmental improvement than providing direct incentives. Consumption penalties such as a carbon tax that directly affects all or many sectors of an economy should lead to a greater number of organizations implementing processes and strategies in an attempt to minimize potentially higher operating costs and barriers, so that they can stay competitive in an increasingly environmentally conscious society.

A balance between providing incentives and penalties to encourage greater uptake of more environmentally preferred practices is more than likely to be the best option. Revenue collected from a financial-based penalty system can then be used to fund large-scale environmental programmes, including providing incentives in the form of rebates and grants.

6.1.4.3 Implications of a penalty-based system for improving the environmental performance of the built environment

Despite the environmental benefits that a penalty on poor environmental performance, such as a carbon tax, should provide, there may be a range of subsequent consequences on the construction industry and others that must be balanced against these environmental benefits. One of the most obvious of these is the ramification of an increase in construction costs on society, including housing affordability, and the viability of the construction industry, which represents a significant proportion of GDP in many countries of the world.

As an example of the potential implications of a carbon tax on the cost of building construction, the carbon dioxide emissions associated with the residential building case study used in Chapter 4 have been used. The total emissions associated with the building's construction are 239.6 t CO_2-e. At a carbon tax rate of A\$40 per tonne of CO_2, this equates to an additional A\$9,584 on top of the initial construction cost. This represents a 4.8 per cent cost increase based on the approximate construction cost for the house (A\$200,005). This may appear a small increase in the cost of construction, but considering the housing affordability issues that are already being faced by many people, especially in the larger cities of the world, this will only further exacerbate these problems. An increase in the cost of construction will inevitably be passed on to developers and purchasers, affecting the affordability of housing, particularly for the most vulnerable. This, together with the typical higher costs associated with improving housing environmental performance, may in turn reduce demand for construction unless existing affordability issues are better addressed or we find low-cost solutions to improve housing environmental performance.

6.2 Where to from here?

When it comes to reducing the environmental impacts of the built environment, to a large extent we are stuck with many of the poor decisions of the past, those that gave little thought to resource efficiency and depletion, human health, wastage and environmental degradation. Many of the existing buildings in many cities of the world were designed in a way that does not respond to their surrounding environment and consequently has resulted in great demand for (often fossil-fuel-based) energy to operate artificial heating, cooling and lighting systems and other services. Replacing all of the poor-performing existing buildings and infrastructure systems that form the built environment is not an option, not just from a logistical point of view but also from the point of view that we should be maximizing the value of the resources embodied in these assets. In some individual cases, demolition may be an option where for various reasons (safety, functional, economic, technological or social) it is not feasible to use these assets any longer, or where the continuing impacts associated with maintenance and operation far outweigh the resource value embodied in that asset and the impacts associated with its replacement.

Whilst significant inroads have been made, particularly over the past decade, in reducing some of the (particularly operational) impacts of the built environment, many of the benefits achieved have been offset by rapid growth in construction resulting from increasing global urbanization. This growth is not expected to abate in the near future either, as it is being further enhanced by a growing global population and rising standards of living. So, where does this leave us? It means, despite a rapidly expanding global population, we (individuals, designers, developers, owners, occupiers and governments) must strive to do everything possible to reduce the environmental impact of existing built environment assets (maximizing their intrinsic resource value and improving their operational efficiency), as well as procuring new buildings (where they are necessary) that have a minimal environmental impact across their entire life cycle. These are the fundamental principles that must drive our future built environment and may at first appear rather simple in nature; however the previous chapters have shown just how complex adhering to these principles can and will actually be. New materials will need to be developed that are renewable or infinitely recyclable; reliable renewable energy systems will be needed to replace current fossil-fuel-based systems; new production processes and systems will need to be developed to significantly improve resource efficiency; and a change in current attitudes and behaviours will also be essential to support and drive the substantial transformation needed.

Environmental improvement must also be balanced with the other important priorities of society: maintaining or improving current standards of living; providing employment, shelter and food; and giving people the freedom to prosper. We must remember though that none of this will be possible or necessary without a planet that is capable of sustaining human life and so the protection of the environment must surely be our number one priority.

6.2.1 Implications of the limitations of current assessment techniques for achieving optimal outcomes in the built environment

Whilst environmental assessment tools are a necessary means for identifying potential environmental impacts and improving current environmental performance across the built environment, the significant limitations of many of these current tools and approaches may severely limit our ability to achieve these ambitious goals. The issues associated with access to environmental data and the time and money needed to thoroughly address the imperative environmental issues constrain built environment professionals in understanding the true and full environmental implications of their choices and making the most environmentally preferred decisions. The international LCA community has a key role to play in improving LCA methodologies and the application of LCA within the built environment, including building awareness amongst and educating those involved in the built environment and improving access to reliable and comprehensive LCA data and tools.

Building regulations must address broad environmental issues and not only focus on just limited aspects of a building's environmental performance, such as their energy consumption related to their thermal performance. They must encourage and enforce the minimization of broader life cycle impacts, particularly those indirect impacts attributable to the resources needed to build our buildings, roads, bridges and infrastructure networks. There must also be incentives for those involved in designing, constructing and managing the built environment to strive for minimal environmental impact and go beyond minimum compliance or hurdle requirements that are often seen as more of a barrier or impediment to traditional design practices. A more holistic and considered approach to the design and management of the built environment must be encouraged.

Also, an additional question still remains, that environmental assessment in itself is unable to answer; is a particular option or solution actually *sustainable*? An environmental assessment may show the extent of the environmental impacts associated with a product or process or which option results in the lowest environmental impacts, but environmental assessment does not show whether any of the options considered are actually sustainable. A deeper understanding of rates of consumption and regeneration, the Earth's capacity to recover from environmental damage and the mechanisms by which the natural systems of the Earth operates is necessary in order to determine whether certain processes or activities are ultimately sustainable in the long term.

6.2.2 Future development of life cycle assessment

It was shown in Chapter 5 that even though the best intentions may exist for reducing environmental impacts, limitations and errors associated with assumptions, together with data completeness, accuracy and relevance and assessment approaches can sometimes work against these intentions, in ways that are often unknown to the decision makers. This is particularly the case with LCA, but despite this it is still currently the best method of quantifying environmental impacts and essential for reducing the impacts occurring across the entire life cycle of any product. In order to alleviate some of these issues, the possible inaccuracies and limitations of LCA studies must be reduced as much as possible. This could be done by:

- collecting and regularly up-dating process data covering a broader range of producers, manufacturers and industries

- using a common approach for assessing indirect environmental impacts to better facilitate more reliable and useful comparisons between alternative materials, products and processes
- developing comparable environmental data using a standard methodology covering the complete system boundary and all inputs and outputs associated with any product or process
- simplifying the assessment process, without affecting the accuracy or relevance of the findings, to minimize costs and time needed for conducting an LCA.

6.2.3 Integration of life cycle assessment into industry

For optimal environmental outcomes to be achieved, the knowledge and tools necessary for environmental assessment and improvement must be fully integrated into the practices of the industries and organizations responsible for designing, constructing and operating the built environment. This may, and most likely will, involve changes to the current practices of these industries and organizations, away from many of the practices (and their resource-intensive nature) that are responsible for the current environmental situation, to those that reflect more responsible and more efficient resource consumption and closer monitoring and management of long-term environmental impacts.

Early-stage environmental assessment is essential to maximize the opportunities for environmental improvement and minimize the costs and other challenges of realizing these opportunities. Tools that are developed for use by industry must provide the most comprehensive and reliable information to help support decision making at the earliest stage of the decision-making process (whether these decisions relate to the design or improvement of processes, products or even whole buildings). In addition, these tools must not impose significant additional costs or time on the design or improvement processes. Design professionals are already submerged in a vast and complex range of regulatory and professional responsibilities and no matter how much of an essential aspect of the design process environmental considerations must be, they do pose an even greater demand on professionals' limited time and resources. For this reason, access to and the use of trained practitioners and consultants is essential. Architects, building designers and managers should work with LCA professionals to assess the environmental impacts of possible design solutions to gain the information they need to realize the full potential of environmental improvement opportunities whilst minimizing any further unwanted imposts on the design process.

Further awareness of the benefits that LCA can provide to industry and individual organizations must also be developed. The LCA community (including practitioners and academics) must play a key role in this process to ensure that the widespread uptake of LCA within the industries responsible for the design, construction and operation of the built environment is achieved. Without an awareness of the significant benefits that the use of LCA can provide (for example, reputational, financial, risk minimization, environmental etc.), how can these organizations be expected to take on board this additional responsibility?

6.3 Conclusion

In order to achieve the ambitious environmental goals needed to avert further degradation of the planet, all of the stakeholders in the built environment must work together. It is unrealistic, and unlikely to lead to the necessary and essential environmental outcomes, to expect any one stakeholder to take full responsibility for the considerable challenges with which we are currently faced.

Industry, including the manufacturers of building materials and components, must improve current manufacturing practices to reduce resource consumption and the production of waste and other pollutants. They must look for alternative, more environmentally friendly ways of operating and even develop new materials that have a lower environmental impact. Those companies that better meet the increasing environmental expectations and desires of clients will also benefit financially by taking a leading role in the move towards a more environmentally conscious built environment. Many companies are now beginning to see the benefits of marketing more environmentally friendly materials and products to a client base that is increasingly demanding them.

Designers must better educate themselves in order to make the most reliable and informed environmental design decisions, ensuring that they provide clients with the best possible environmental outcomes within their projects. They must specify materials and systems and design buildings to maximize resource efficiencies over the long term. Where uncertainty still exists, designers should seek advice and assistance to better understand the full environmental consequences of their decisions.

Clients of built asset projects must demand best-practice environmental practices and outcomes for their projects. This is particularly crucial as client involvement, demands and expectations can have a significant influence over the way in which industry operates. After all, it is clients' money that keeps many of these organizations in business.

Governments must put in place policies that help to improve and encourage better environmental approaches to the design and management of the built environment. This may come in various forms, through building regulations, environmental controls and standards or financial incentives. They must also play a key role in educating consumers, manufacturers, designers and developers on how each of these stakeholders can achieve improved environmental outcomes in their everyday decision making. Also, as with any newly developing field, further research is an essential element for the progression of environmental change. Funding schemes to explore, test and develop new technologies, materials and processes must be established by governments to encourage the development of innovative solutions to key environmental problems.

Each stakeholder group must also have leaders that continually strive to improve best-practice environmental performance, demonstrating what is possible and how this might be achieved. Through this leadership, other organizations will follow in order to maintain market share in an increasingly competitive marketplace, where environmental concerns of consumers are becoming an increasingly important factor in their purchasing decisions.

Whether it is climate change – the pollution of the air we breathe, the water we drink or the land we rely on for growing our food – or the depletion of non-renewable resources, the urgency to minimize the impacts of human activity on the environment is only becoming more critical as the world's population expands and standards of living increase. No longer can we hope that someone else will provide the solution: we must all act now by striving to improve the environmental performance of everything we do. This book has shown how the LCA tool is an essential component of the solution towards a more sustainable world, how it can be used to improve the environmental performance of the built environment and what we must do now to achieve these ambitious goals. But using LCA is only one small part of the solution. We must all act on our desire and responsibility to afford future generations the same or better standards of living than what we enjoy today without destroying the very resources on which human survival relies.

Appendix A

Table A.1 Top 100 energy pathways from the Australian *Residential building* sector, based on 1996-97 input-output tables

No.	Stage 1	Stage 2	Stage 3	Stage 4	DER (GJ/A$1000)
1	Ceramic products				0.6629
2	Cement, lime and concrete slurry				0.4018
3	DIRECT				0.3490
4	Road transport				0.2335
5	Iron and steel				0.2160
6	Structural metal products	Iron and steel			0.1817
7	Other non-metallic mineral products				0.1643
8	Household appliances	Iron and steel			0.0924
9	Cement, lime and concrete slurry	Road transport			0.0922
10	Residential building	Ceramic products			0.0691
11	Plastic products	Basic chemicals			0.0548
12	Fabricated metal products	Iron and steel			0.0515
13	Iron and steel	Iron and steel			0.0491

No.	Stage 1	Stage 2	Stage 3	Stage 4	DER (GJ/A$1000)
14	Cement, lime and concrete slurry	Cement, lime and concrete slurry			0.0436
15	Residential building	Cement, lime and concrete slurry			0.0419
16	Structural metal products	Iron and steel	Iron and steel		0.0413
17	Ceramic products	Other non-metallic mineral products			0.0411
18	Wholesale trade				0.0370
19	Residential building				0.0364
20	Structural metal products	Basic non-ferrous metal and products			0.0360
21	Paints	Basic chemicals			0.0353
22	Plaster and other concrete products	Cement, lime and concrete slurry			0.0352
23	Plaster and other concrete products	Road transport			0.0345
24	Ceramic products	Road transport			0.0326
25	Sawmill products	Road transport			0.0325
26	Other wood products				0.0317
27	Other wood products	Road transport			0.0293

No.	Stage 1	Stage 2	Stage 3	Stage 4	DER (GJ/A$1000)
28	Other mining				0.0285
29	Sawmill products				0.0276
30	Road transport	Road transport			0.0260
31	Plaster and other concrete products				0.0252
32	Residential building	Road transport			0.0244
33	Sheet metal products	Iron and steel			0.0236
34	Other wood products	Basic chemicals			0.0230
35	Residential building	Iron and steel			0.0225
36	Structural metal products	Structural metal products	Iron and steel		0.0213
37	Household appliances	Iron and steel	Iron and steel		0.0210
38	Cement, lime and concrete slurry	Other mining			0.0194
39	Wholesale trade	Road transport			0.0193
40	Residential building	Structural metal products	Iron and steel		0.0189
41	Structural metal products				0.0181
42	Residential building	Other non-metallic mineral products			0.0171
43	Other non-metallic mineral products	Road transport			0.0169

No.	Stage 1	Stage 2	Stage 3	Stage 4	DER (GJ/A$1000)
44	Agricultural, mining and const. machinery	Iron and steel			0.0152
45	Plastic products	Basic chemicals	Basic chemicals		0.0151
46	Fabricated metal products	Basic non-ferrous metal and products			0.0149
47	Other machinery and equipment	Iron and steel			0.0149
48	Sawmill products	Forestry and logging			0.0147
49	Plaster and other concrete products	Iron and steel			0.0146
50	Wholesale trade	Air and space transport			0.0130
51	Fabricated metal products				0.0118
52	Other non-metallic mineral products	Basic chemicals			0.0117
53	Fabricated metal products	Iron and steel	Iron and steel		0.0117
54	Textile products				0.0112
55	Iron and steel	Iron and steel	Iron and steel		0.0112
56	Other electrical equipment	Basic non-ferrous metal and products			0.0108
57	Basic non-ferrous metal and products				0.0103
58	Cement, lime and concrete slurry	Road transport	Road transport		0.0103
59	Structural metal products	Road transport			0.0102

No.	Stage 1	Stage 2	Stage 3	Stage 4	DER (GJ/A$1000)
60	Cement, lime and concrete slurry	Cement, lime and concrete slurry	Road transport		0.0100
61	Sheet metal products	Basic non-ferrous metal and products			0.0100
62	Air and space transport				0.0099
63	Paints	Basic chemicals	Basic chemicals		0.0097
64	Residential building	Household appliances	Iron and steel		0.0096
65	Residential building	Cement, lime and concrete slurry	Road transport		0.0096
66	Other wood products	Sawmill products	Road transport		0.0096
67	Structural metal products	Iron and steel	Iron and steel	Iron and steel	0.0094
68	Fabricated metal products	Basic chemicals			0.0090
69	Structural metal products	Basic non-ferrous metal and products	Basic non-ferrous metal and products		0.0090
70	Glass and glass products	Glass and glass products			0.0089
71	Structural metal products	Glass and glass products			0.0088
72	Household appliances	Basic chemicals			0.0084
73	Other wood products	Pulp, paper and paperboard			0.0083

No.	Stage 1	Stage 2	Stage 3	Stage 4	DER (GJ/A$1000)
74	Other wood products	Sawmill products			0.0081
75	Plaster and other concrete products	Cement, lime and concrete slurry	Road transport		0.0081
76	Household appliances	Other electrical equipment	Basic non-ferrous metal and products		0.0075
77	Household appliances	Household appliances	Iron and steel		0.0073
78	Residential building	Residential building	Ceramic products		0.0072
79	Other mining	Road transport			0.0072
80	Plaster and other concrete products	Other mining			0.0067
81	Sawmill products	Pulp, paper and paperboard			0.0064
82	Ceramic products	Ceramic products			0.0063
83	Other wood products	Basic chemicals	Basic chemicals		0.0063
84	Household appliances	Road transport			0.0063
85	Other electrical equipment	Basic chemicals			0.0063
86	Other non-metallic mineral products	Cement, lime and concrete slurry			0.0062
87	Other chemical products	Basic chemicals			0.0062
88	Structural metal products	Basic chemicals			0.0058

No.	Stage 1	Stage 2	Stage 3	Stage 4	DER (GJ/A$1000)
89	Residential building	Plastic products	Basic chemicals		0.0057
90	Ceramic products	Cement, lime and concrete slurry			0.0057
91	Sawmill products	Sawmill products	Road transport		0.0057
92	Household appliances	Basic non-ferrous metal and products			0.0056
93	Basic chemicals				0.0055
94	Other non-metallic mineral products	Other mining			0.0055
95	Structural metal products	Fabricated metal products	Iron and steel		0.0055
96	Iron and steel	Basic non-ferrous metal and products			0.0055
97	Sheet metal products	Iron and steel	Iron and steel		0.0054
98	Residential building	Fabricated metal products	Iron and steel		0.0054
99	Other property services				0.0053
100	Residential building	Iron and steel	Iron and steel		0.0051

Total of top 100 pathways   3.9795

Proportion of sector total energy requirement   37%

Table A.2 Embodied energy and water material coefficients

Material	Unit	Embodied energy coefficient (GJ/unit)	Embodied water coefficient (kL/unit)
Aluminium			
virgin	t	252.6	1084.4
reflective foil	m²	0.137	0.588
Asphalt	m³	3.08	
Bitumen	m³	48.39	11.86
Carpet			
wool	m²	0.741	2.18
nylon	m²	0.683	1.58
Ceramics			
clay bricks (110 mm)	m²	0.56	0.672
ceramic tiles	m²	0.293	1.12
terracotta roof tiles (20 mm)	m²	0.986	1.52
Concrete			
5 MPa concrete	m³	2.79	7.48
15 MPa concrete	m³	4.03	10.13
20 MPa concrete	m³	4.44	10.98
25 MPa concrete	m³	5.01	11.80
30 MPa concrete	m³	5.44	12.29
32 MPa concrete	m³	5.81	13.10
40 MPa concrete	m³	6.75	14.62
aerated concrete (200 mm)	m²	0.495	1.64
cement	t	16.96	29.91
concrete block, hollow (200 mm)	m²	0.805	2.67
concrete roof tile (20 mm)	m²	0.251	0.909
fibre cement sheet (4.5 mm)	m²	0.235	0.745
fibre cement sheet (6 mm)	m²	0.288	0.883
mortar	m³	2.00	10.55
precast	m³	4.44	10.98
Glass			
clear float glass (4 mm)	m²	1.73	3.42

Material	Unit	Embodied energy coefficient (GJ/unit)	Embodied water coefficient (kL/unit)
glass fibre	m^3	432.1	855.3
toughened glass (6 mm)	m^2	3.66	8.24
toughened glass (12 mm)	m^2	7.31	16.49
Insulation			
expanded polystyrene insulation (50 mm)	m^2	0.361	0.734
fibreglass insulation (80 mm)	m^2	0.183	0.376
fibreglass insulation (100 mm)	m^2	0.217	0.429
fibreglass insulation (160 mm)	m^2	0.348	0.687
Paint			
oil-based paint	m^2	0.101	0.219
water-based paint	m^2	0.096	0.212
Plasterboard			
plasterboard (10 mm)	m^2	0.207	0.627
plasterboard (13 mm)	m^2	0.232	0.682
Plastics			
general (PVC)	t	156.9	366.4
laminate (1 mm)	m^2	0.200	0.476
plastic membrane (1 mm)	m^2	0.514	1.40
polyester	t	156.9	366.4
polystyrene	m^3	7.04	14.16
PVC water pipe (20 mm)	m	0.212	0.452
UPVC pipe 100	m	0.266	0.568
UPVC pipe 100 (slotted)	m	0.208	0.443
vinyl flooring (2 mm)	m^2	0.661	1.72
Sand and stone			
granite	t	0.087	
sand	m^3	0.617	3.57
screenings	m^3	0.691	4.43
Steel			
COLORBOND® steel decking	m^2	0.933	1.34

Material	Unit	Embodied energy coefficient (GJ/unit)	Embodied water coefficient (kL/unit)
reinforcement	t	85.46	98.64
stainless steel	t	445.2	649.6
steel	t	85.46	98.64
steel decking	m^2	0.796	1.12
Timber			
hardwood	m^3	21.33	16.28
MDF/particleboard	m^3	30.35	85.59
softwood	m^3	10.93	20.14
Oil	m^3	34	
Other metals			
Copper	t	378.9	1188.4

Note: These energy and water coefficients were compiled by Treloar and Crawford (2010) based on an energy and water-based input-output model developed by Prof. Manfred Lenzen at the University of Sydney and Australian process data compiled by Grant (2002).

Table A.3 Initial embodied energy calculation of residential building case study

Element	Item	Material	Unit	Quantity	Embodied energy coefficient (GJ/unit)	Embodied energy (GJ)
Substructure	Concrete slab	20 MPa concrete	m³	59.54	4.44	264.13
	Waffle pods	Polystyrene	m³	46.72	7.04	328.76
	Reinforcement	Steel	t	5.44	85.46	464.61
	Membrane	Plastic membrane (1 mm)	m²	326.7	0.514	168.08
	Gravel to base	Screenings	m³	6.5	0.691	4.49
	Sewerage and stormwater drains	UPVC pipe 100	m	157.5	0.266	41.97
Walls	Damp proof course	Plastic membrane (1 mm)	m²	28.35	0.514	14.59
	Brick ties	Steel	t	0.0162	85.46	1.39
	Softwood framing	Softwood	m³	8.27	10.93	90.37
	Hardwood framing	Hardwood	m³	0.488	21.33	10.41
	Fixings	Steel	t	0.0087	85.46	0.75
	Bracing	Steel	t	0.028	85.46	2.39
	Post support	Steel	t	0.0042	85.46	0.36
	Masonite brace board	MDF/particleboard	m³	0.233	30.35	7.06
	Steel bolts and washers	Steel	t	0.0061	85.46	0.52
	Fibre cement sheet infills	Fibre cement sheet (4.5 mm)	m²	4.06	0.235	0.96

Element	Item	Material	Unit	Quantity	Embodied energy coefficient (GJ/unit)	Embodied energy (GJ)
	Glulam beam (4.5 m)	Softwood	m³	0.138	10.93	1.51
	LVL beam 300 x 45	Softwood	m³	0.302	10.93	3.30
	Brick cladding, sills and piers	Clay bricks	m²	97.21	0.56	54.48
	Brick sand	Sand	m³	13.42	0.617	8.29
	Steel lintels	Steel	t	0.304	85.46	25.96
	Glasswool wall batts (R2.0)	Fibreglass insulation (80 mm)	m²	175.53	0.183	32.10
	Sisalation to external walls	Aluminium reflective foil	m²	175.53	0.137	24.03
	Plasterboard cladding and cornices	Plasterboard (10 mm)	m²	717.31	0.207	148.31
Windows	Sliding aluminium 1200 x 1810	Aluminium	t	0.011	252.6	2.74
		Clear float glass (4 mm)	m²	2.17	1.73	3.75
	Sliding aluminium 1543 x 1450	Aluminium	t	0.011	252.6	2.83
		Clear float glass (4 mm)	m²	2.24	1.73	3.87
	Aluminium 514 x 1450	Aluminium	t	0.0075	252.6	1.88
		Clear float glass (4 mm)	m²	1.49	1.73	2.58
	Aluminium 514 x 1810	Aluminium	t	0.0093	252.6	2.35
		Clear float glass (4 mm)	m²	1.86	1.73	3.22
	Aluminium 1543 x 1810 - double glazed	Aluminium	t	0.014	252.6	3.53

Element	Item	Material	Unit	Quantity	Embodied energy coefficient (GJ/unit)	Embodied energy (GJ)
		Clear float glass (4 mm)	m²	5.59	1.73	9.65
	Aluminium 1800 x 2410 - double glazed	Aluminium	t	0.022	252.6	5.48
		Clear float glass (4 mm)	m²	8.68	1.73	14.99
	Aluminium 1800 x 2704 - double glazed	Aluminium	t	0.024	252.6	6.15
		Clear float glass (4 mm)	m²	9.73	1.73	16.82
	Aluminium 1800 x 610 - mock casement	Aluminium	t	0.011	252.6	2.77
		Clear float glass (4 mm)	m²	2.20	1.73	3.80
	Aluminium 1800 x 1210 - mock casement	Aluminium	t	0.022	252.6	5.50
		Clear float glass (4 mm)	m²	4.36	1.73	7.53
	Sliding aluminium door 2110 x 1450	Aluminium	t	0.015	252.6	3.86
		Clear float glass (4 mm)	m²	3.06	1.73	5.29
	Sliding aluminium door 2110 x 3250	Aluminium	t	0.103	252.6	25.98
		Clear float glass (4 mm)	m²	20.57	1.73	35.56
Roof	Roof trusses	Softwood	m³	11.02	10.93	120.40

Element	Item	Material	Unit	Quantity	Embodied energy coefficient (GJ/unit)	Embodied energy (GJ)
	Valley irons	Steel	t	0.086	85.46	7.35
	Quad gutter, fascia, downpipes	COLORBOND® steel decking	t	0.41	0.933	0.38
	Barge fascia & dry soaker	COLORBOND® steel decking	t	0.0076	0.933	0.01
	Hip capping	Concrete roof tile (20 mm)	m²	12.16	0.251	3.06
	Ridge capping	Concrete roof tile (20 mm)	m²	4.20	0.251	1.06
	Dry valley	Hardwood	m³	0.183	21.33	3.89
	Dry gable	Hardwood	m³	0.065	21.33	1.38
	Concrete roof tiles	Concrete roof tile (20 mm)	m²	415.61	0.251	104.46
	Sarking to long length rafters	Aluminium reflective foil	m²	33.49	0.137	4.59
	Glasswool ceiling batts (R4.0)	Fibreglass ins. (160 mm)	m²	266.38	0.348	92.58
	Additional downpipe	COLORBOND® steel decking	t	0.027	0.933	0.02
	Zincalume lear gutter with apron	Steel	t	0.021	85.46	1.83
	Zincalume flashing for lightweight cladding	Steel	t	0.03	85.46	2.54
	Plasterboard to ceilings	Plasterboard (10 mm)	m²	295.99	0.207	61.2
Doors	Flush panel doors	MDF/particleboard	m³	0.862	30.35	26.15
	Jambs	MDF/particleboard	m³	0.21	30.35	6.37
	Frames	Hardwood	m³	0.052	21.33	1.11

Element	Item	Material	Unit	Quantity	Embodied energy coefficient (GJ/unit)	Embodied energy (GJ)
	Door hardware	Steel	t	0.0052	85.46	0.44
Internal finishes	Wall tiles	Ceramic tiles	m²	33.24	0.293	9.75
	Floor tiles	Ceramic tiles	m²	26.71	0.293	7.83
	Aluminium tiling angle - 1.0 m long	Aluminium	t	0.00085	252.6	0.21
	Aluminium L-shaped angle tile trim	Aluminium	t	0.001	252.6	0.26
	Carpet	Nylon carpet	m²	252.53	0.683	172.54
	Paint – walls	Water-based paint	m²	985.68	0.096	95.01
	Paint – ceilings	Water-based paint	m²	291.54	0.096	28.10
	Paint - doors, jambs, architraves, reveals	Water-based paint	m²	81.49	0.096	7.85
	Cladding for bath hobs	MDF/particleboard	m³	0.115	30.35	3.49
External finishes	Fibre cement sheet infills	Fibre cement sheet (4.5 mm)	m²	61.24	0.235	14.42
	Timber joinery	Hardwood	m³	0.042	21.33	0.89
	Timber joinery	Softwood	m³	0.581	10.93	6.34
	PVC joint strip	Plastic	t	0.0009	156.9	0.14
	Masonite packing strips	MDF/particleboard	m³	0.015	30.35	0.46
	Exterior grade plywood (15 mm)	MDF/particleboard	m³	0.045	30.35	1.38
	Timber battens	Softwood	m³	0.163	10.93	1.79

Element	Item	Material	Unit	Quantity	Embodied energy coefficient (GJ/unit)	Embodied energy (GJ)
	Paint - eaves, lintels, doors, infills, beams	Water-based paint	m²	59.75	0.096	5.76
	Paint - battens, entry frame	Oil-based paint	m²	7.64	0.101	0.77
Internal fitout	Semi-frameless shower screen	Toughened glass (6 mm)	m²	3.708	3.66	13.56
		Aluminium	t	0.00027	252.6	0.068
	Semi-frameless shower screen	Toughened glass (6 mm)	m²	2.472	3.66	9.04
		Aluminium	t	0.00036	252.6	0.091
	Sliding robe door – mirrored	Clear float glass (4 mm)	m²	4.47	1.73	7.72
		Aluminium	t	0.00063	252.6	0.16
	Sliding robe door – mirrored	Clear float glass (4 mm)	m²	5.11	1.73	8.82
		Aluminium	t	0.00072	252.6	0.18
	Sliding robe door – mirrored	Clear float glass (4 mm)	m²	5.74	1.73	9.93
		Aluminium	t	0.00081	252.6	0.2
	Sliding robe door – mirrored	Clear float glass (4 mm)	m²	9.78	1.73	16.91
		Aluminium	t	0.0014	252.6	0.35
	White melamine shelving	MDF/particleboard	m³	0.199	30.35	6.03
		Laminate (1 mm)	m²	12.87	0.20	2.58
	Hanging rod – 25 mm diameter	Steel	t	0.06	85.46	5.12

Element	Item	Material	Unit	Quantity	Embodied energy coefficient (GJ/unit)	Embodied energy (GJ)
	Support bracket – white	Plastic	t	0.0007	156.9	0.11
	Melamine shelving (16 mm)	MDF/particleboard	m³	0.097	30.35	2.94
		Laminate (1 mm)	m²	6.26	0.20	1.26
	Polished edge mirrors (4 mm)	Clear float glass (4 mm)	m²	4.83	1.73	8.34
	Taps and shower heads	Steel	t	0.0134	85.46	1.14
	Vitreous china toilet suite	Ceramic toilet suite	no.	2	9.91	19.83
	Door handles	Stainless steel	t	0.0065	445.2	2.9
	Sectional overhead garage door	Steel	t	0.00013	85.46	0.01
	MDF skirting 67 x 12 mm	MDF/particleboard	m³	0.499	30.35	15.14
	Masonite packing strips	MDF/particleboard	m³	0.003	30.35	0.09
	Meranti 19 x 19 mm	Hardwood	m³	0.015	21.33	0.32
	Laundry sink (45 litre)	Laundry sink (45 litre)	no.	1	11.08	11.08
	Stainless steel sink (1120 mm)	Stainless steel sink	no.	1	13.76	13.76
	Vanity basin	Ceramic basin	no.	4	10.73	42.92
	Mirror back board (16 mm)	MDF/particleboard	m³	0.049	30.35	1.48
	Joinery - cupboards	MDF/particleboard	m³	1.05	30.35	31.87
	Stove	Stove	no.	1	16.10	16.10

Element	Item	Material	Unit	Quantity	Embodied energy coefficient (GJ/unit)	Embodied energy (GJ)
	Ducted heating system	Ducted heating system	no.	1	60.00	60.00
	Gas boosted solar HWS, 200L tank	Solar hot water service	no.	1	57.75	57.75
	Canopy rangehood	Rangehood	no.	1	12.88	12.88
	Flow & return line for solar HWS	Copper	t	0.013	378.9	4.89
	Hinges 85 mm with screws	Steel	t	0.0007	85.46	0.06
	Mortar for baths and shower bases	Mortar	t	0.078	2.0	0.16
	Zinc angle 40 x 40 x 2400	Aluminium	t	0.021	252.6	5.22
	Shower base 900 x 900	Plastic	t	0.0081	156.9	1.27
	Bath 1675 mm - white acrylic	Acrylic bath (1680 long)	no.	2	5.52	11.05
	Water pipes	Plastic	t	0.02	156.9	3.10
	Aluminium window reveals & architraves	MDF/particleboard	m³	0.271	30.35	8.24
Electrical	Weatherproof single socket outlet	Plastic	t	0.0001	156.9	0.02
	Weatherproof flood light	Plastic	t	0.0001	156.9	0.02
	Light points	Plastic	t	0.0019	156.9	0.299
	Power outlets	Plastic	t	0.0013	156.9	0.21
	Exhaust fans (250 mm)	Plastic	t	0.0041	156.9	0.65
	TV point with 6 m of co-axial cable	Plastic	t	0.00034	156.9	0.05

Element	Item	Material	Unit	Quantity	Embodied energy coefficient (GJ/unit)	Embodied energy (GJ)
		Copper	t	0.0016	378.9	0.59
	Conical light shades	Plastic	t	0.0016	156.9	0.24
	Draft stopper	Plastic	t	0.0062	156.9	0.97
				Total initial embodied energy		3,082

Table A.4 Input-output pathways of *Residential building* sector representing quantified materials for residential building case study

Input-output pathway	Materials	Elements	TER (GJ/A$1000)
Basic chemicals	Polystyrene	Slab (waffle pods)	0.0100
Basic non-ferrous metal and products	Aluminium reflective foil, aluminium, copper	Sisalation, tile trim, wardrobe and shower frames, water pipes	0.0256
Cement, lime and concrete slurry	25 MPa concrete, mortar	Slab, brick walls	0.8988
Ceramic products	Clay bricks, ceramic tiles, ceramics	External walls, wall and floor finishes, toilet suite, basin	1.0940
Fabricated metal products	Steel	Fixings, bracing, bolts, washers, door handles, taps, garage door, sinks, hinges	0.2555
Glass and glass products	Toughened glass (6 mm), clear float glass (4 mm)	Shower screens, mirrors	0.0226
Household appliances	Various	Solar hot water system, rangehood, stove	0.4508
Iron and steel	Steel	Bracing	0.4222
Other electrical equipment	Copper	Co-axial cabling	0.0848
Other machinery and equipment	Various	Ducted heating system	0.0518
Other mining	Screenings, sand	Slab base, brick walls	0.0770
Other non-metallic mineral products	Fibreglass insulation (various thicknesses)	Wall and ceiling insulation	0.3490
Other wood products	MDF/particleboard, softwood	Bracing, beams, roof trusses, doors, jambs, frames, shelving, skirting, mirror backing, joinery, architraves, reveals	0.5704

Input-output pathway	Materials	Elements	TER (GJ/A$1000)
Paints	Water-based paint, oil-based paint	Wall, ceiling, door, jamb, architrave, reveal, eave, lintel, beam and batten finishes	0.1010
Plaster and other concrete products	Fibre cement sheet (4.5 mm), plasterboard (10 mm), concrete roof tile (20 mm)	External wall infills, internal wall and ceiling cladding, cornice, roof cladding	0.3361
Plastic products	Plastic membrane (1 mm), UPVC pipe 100, plastic	Slab, drains, damp proof course, shower base, bath, water pipes, light points, power outlets, exhaust fans, light shade, wiring	0.1675
Sawmill products	Softwood, hardwood	Wall framing, roof framing, external joinery, battens	0.2722
Sheet metal products	Steel	Brick ties, bracing, roof valleys, fascias, gutters, downpipes, flashings	0.0942
Structural metal products	Steel, windows (aluminium, clear float glass (4 mm))	Slab reinforcement, lintels, aluminium windows	0.7497
Textile products	Nylon carpet	Floor covering	0.0410
		Total	6.074

Appendix B

Table B.1 Initial embodied energy calculation of concrete-framed building

Element	Material	Unit	Quantity	Embodied energy coefficient (GJ/unit)	Embodied energy (GJ)
Footings	40 MPa concrete	m³	9,527	6.75	64,279
	Steel reinforcement	t	755	85.46	64,525
Beams	30 MPa concrete	m³	8,915	5.44	48,524
	Steel reinforcement	t	850	85.46	72,644
Columns	40 MPa concrete	m³	1,820	6.75	12,280
	Steel reinforcement	t	292	85.46	24,955
Floor slabs	30 MPa concrete	m³	9,276	5.44	50,489
	Steel reinforcement	t	587	85.46	50,167
Shear walls	30 MPa concrete	m³	8,795	5.44	47,871
	Steel reinforcement	t	697	85.46	59,568
Staircase	30 MPa concrete	m³	163	5.44	887
	Steel reinforcement	t	13	85.46	1,111
Roof slab	30 MPa concrete	m³	410	5.44	2,232
	Steel reinforcement	t	32	85.46	2,735
Façade	Clear float glass (4 mm)	m²	126,250	1.73	218,160
	Aluminium framing	t	15	252.6	3,789
				Total initial embodied energy	724,215

Table B.2 Initial embodied energy calculation of steel-framed building

Element	Material	Unit	Quantity	Embodied energy coefficient (GJ/unit)	Embodied energy (GJ)
Footings	40 MPa concrete	m³	9,527	6.75	64,279
	Steel reinforcement	t	755	85.46	64,525
Beams	Steel, structural	t	3,722	85.46	318,093
Columns	Steel, structural	t	2,083	85.46	178,019
Floor slabs	30 MPa concrete	m³	10,106	5.44	55,007
	Steel reinforcement	t	640	85.46	54,696
Shear walls	30 MPa concrete	m³	8,795	5.44	47,871
	Steel reinforcement	t	697	85.46	59,568
Staircase	30 MPa concrete	m³	163	5.44	887
	Steel reinforcement	t	13	85.46	1,111
Roof slab	30 MPa concrete	m³	241	5.44	1,312
	Steel reinforcement	t	15	85.46	1,282
	Steel, structural (beams)	t	74	85.46	6,324
Façade	Clear float glass (4 mm)	m²	126,250	1.73	218,160
	Aluminium framing	t	15	252.6	3,789
				Total initial embodied energy	1,074,923

Table B.3 Initial embodied energy calculation of commercial office building

Element	Item	Material	Unit	Quantity	Embodied energy coefficient (GJ/unit)	Embodied energy (GJ)
Substructure	25 MPa reinforced concrete	25 MPa concrete	m³	400.5	5.01	2,008
		Steel reinforcement	t	41.95	85.46	3,585
		Plastic membrane (1 mm)	m²	85.8	0.514	44.14
	32 MPa reinforced concrete	32 MPa concrete	m³	1,253	5.81	7,278
		Steel reinforcement	t	119.9	85.46	10,247
		Sand	m³	295.1	0.617	182.2
		Plastic membrane	m²	5,901	0.514	3,036
Columns	Steel, structural	Steel	t	196.1	85.46	16,759
		Water-based paint	m²	52.5	0.096	5.06
Upper floors	32 MPa reinforced concrete	32 MPa concrete	m³	996.4	5.81	5,785
		Steel reinforcement	t	95.31	85.46	8,145
		Steel decking	m²	6,380	0.796	5,079
	Steel, structural	Steel	t	240.9	85.46	20,590
Staircases	32 MPa reinforced concrete	32 MPa concrete	m³	19.16	5.81	111.3
		Steel reinforcement	t	2.01	85.46	171.6
	Steel frame to stair	Steel	t	3.39	85.46	289.5

Element	Item	Material	Unit	Quantity	Embodied energy coefficient (GJ/unit)	Embodied energy (GJ)
	Handrails	Steel	t	0.091	85.46	7.81
		Stainless steel	t	0.131	445.2	58.43
		Water-based paint	m²	25.41	0.096	2.45
Roof	Steel, structural	Steel	t	86.41	85.46	7,384
	Entry canopies	Steel	t	0.82	85.46	70.0
		Steel decking	m²	163.8	0.796	130.4
		Aluminium	t	0.18	252.6	44.42
	COLORBOND® metal tray roof	COLORBOND® steel decking	m²	4,506	0.933	4,202
		Steel	t	22.53	85.46	1,925
	Galvanized metal tray roof	Steel decking	m²	1,084	0.796	862.6
		Steel	t	5.42	85.46	463.0
	Glazed roof	Toughened glass (6 mm)	m²	157.6	3.66	576.3
		Aluminium	t	0.8	252.6	202.9
	Gutters, downpipes and flashings	Steel decking	m²	402.2	0.796	320.1
		Aluminium	t	1.51	252.6	381
	Linings to parapets	Aluminium	t	0.68	252.6	172.2
	Tanking	Oil-based paint	m²	260.4	0.101	26.29

Element	Item	Material	Unit	Quantity	Embodied energy coefficient (GJ/unit)	Embodied energy (GJ)
	Roof walkways	Steel decking	m²	212.1	0.796	168.8
		Steel	t	1.06	85.46	90.6
External walls	Precast concrete cladding panels	32 MPa concrete	m³	106.9	5.81	620.4
		Steel reinforcement	t	12.82	85.46	1,095
		Water-based paint	m²	1,336	0.096	128.7
	Aluminium cladding panels	Aluminium	t	3.94	252.6	995.9
		Plasterboard (13 mm)	m²	1,068	0.232	247.9
		Water-based paint	m²	1,068	0.096	102.9
Windows	Frameless solar double glazing	Toughened glass (12 mm)	m²	2,699	7.31	19,738
		Aluminium	t	10.45	252.6	2,639
		Steel	t	14.41	85.46	1,231
	Support frame	Steel	t	26.27	85.46	2,245
	Aluminium curtain wall	Toughened glass (6 mm)	m²	1,235	3.66	4,516
		Aluminium	t	6.29	252.6	1,590
	Aluminium framed windows	Toughened glass (6 mm)	m²	250.3	3.66	915.4
		Aluminium	t	1.28	252.6	322.3
	Aluminium louvres	Aluminium	t	7.25	252.6	1,830

Element	Item	Material	Unit	Quantity	Embodied energy coefficient (GJ/unit)	Embodied energy (GJ)
	Aluminium glazed screens	Toughened glass (6 mm)	m²	137	3.66	501
		Aluminium	t	0.7	252.6	176.4
	Internal frameless glazing	Toughened glass (12 mm)	m²	373.9	7.31	2 734
		Steel	t	3.99	85.46	341.3
External doors	Aluminium doors and frame	Toughened glass (6 mm)	m²	140.1	3.66	512.3
		Aluminium	t	1.48	252.6	374
		Steel	t	0.0034	85.46	0.29
	Timber doors	Hardwood	m³	0.31	21.33	6.59
		Steel	t	0.033	85.46	2.82
		Steel	t	0.0017	85.46	0.14
		Water-based paint	m²	12.6	0.096	1.21
	Fire doors	Hardwood	m³	0.31	21.33	6.59
		Steel	t	0.033	85.46	2.82
		Steel	t	0.002	85.46	0.14
		Water-based paint	m²	12.60	0.096	1.21
Internal walls	Level 1	Plasterboard (10 mm)	m²	10,824	0.207	2,237
		Water-based paint	m²	10,824	0.096	1,043

Element	Item	Material	Unit	Quantity	Embodied energy coefficient (GJ/unit)	Embodied energy (GJ)
		Steel	t	26.98	85.46	2,305
		32 MPa concrete	m³	74.13	5.81	430.4
		Steel reinforcement	t	7.77	85.46	663.7
		Clay bricks	m²	361.2	0.56	202.4
		MDF/particleboard	m³	1.74	30.35	52.68
		Plastic	t	0.09	156.9	14.33
		Aluminium	t	0.12	252.6	30.50
		Softwood	m³	6.90	10.93	75.41
	Level 2	Aerated concrete (200 mm)	m²	255.2	0.495	126.4
		Plasterboard (10 mm)	m²	1,699	0.207	351.3
		32 MPa concrete	m³	9.98	5.81	57.92
		Steel reinforcement	t	1.05	85.46	89.31
		Water-based paint	m²	1,699	0.096	163.8
		MDF/particleboard	m³	1.14	30.35	34.51
		Plastic	t	0.06	156.9	9.39
		Aluminium	t	0.07	252.6	17.24
		Steel	t	6.14	85.46	525

Element	Item	Material	Unit	Quantity	Embodied energy coefficient (GJ/unit)	Embodied energy (GJ)
	Level 3					
		Aerated concrete (200 mm)	m²	255.2	0.495	126.4
		Plasterboard (10 mm)	m²	2,938	0.207	607.4
		32 MPa concrete	m³	9.98	5.81	57.92
		Steel reinforcement	t	1.05	85.46	89.31
		Water-based paint	m²	2,938	0.096	283.2
		Steel	t	6.77	85.46	578.2
		MDF/particleboard	m³	1.40	30.35	42.38
		Plastic	t	0.07	156.9	11.53
		Aluminium	t	0.09	252.6	22.54
		Steel	t	6.58	85.46	562.6
		Plasterboard (10 mm)	m²	2,514	0.207	519.7
		Water-based paint	m²	2,514	0.096	242.3
		Softwood	m³	18.23	10.93	199.2
		MDF/particleboard	m³	0.28	30.35	8.48
		Plastic	t	0.01	156.9	2.31
		Aluminium	t	0.02	252.6	5.30

Element	Item	Material	Unit	Quantity	Embodied energy coefficient (GJ/unit)	Embodied energy (GJ)
Internal screens, windows and glass balustrading	Glazed screens and windows	Toughened glass (6 mm)	m²	1,104	3.66	4,038
		Aluminium	t	7.59	252.6	1,916
		Toughened glass (12 mm)	m²	384.2	7.31	2,810
	Glass balustrade	Toughened glass (6 mm)	m²	242.1	3.66	885.2
		Aluminium	t	1.23	252.6	311.7
Internal doors	Level 1	Hardwood	m³	0.74	21.33	15.82
		Water-based paint	m²	302.4	0.096	29.15
		Stainless steel	t	0.15	445.2	67.31
		Aluminium	t	1.58	252.6	399.2
		Toughened glass (6 mm)	m²	345.1	3.66	1,261
		Toughened glass (12 mm)	m²	2.06	7.31	15.07
		Steel	t	1.80	85.46	154.2
	Level 2	Hardwood	m³	0.32	21.33	6.81
		Water-based paint	m²	130.2	0.096	12.55
		Toughened glass (12 mm)	m²	10.3	7.31	75.34
		Aluminium	t	0.05	252.6	13.26

Element	Item	Material	Unit	Quantity	Embodied energy coefficient (GJ/unit)	Embodied energy (GJ)
		Stainless steel	t	0.07	445.2	28.98
	Level 3	Hardwood	m³	0.33	21.33	7.03
		Water-based paint	m²	134.4	0.096	12.95
		Toughened glass (12 mm)	m²	8.24	7.31	60.27
		Aluminium	t	0.04	252.6	10.61
		Stainless steel	t	0.07	445.2	29.92
	Level 3 directors	Hardwood	m³	0.25	21.33	5.27
		Water-based paint	m²	100.8	0.096	9.72
		Stainless steel	t	0.05	445.2	22.44
		Aluminium	t	0.04	252.6	10.61
		Toughened glass (6 mm)	m²	4.12	3.66	15.07
		Toughened glass (12 mm)	m²	6.18	7.31	45.20
	Level 4	Hardwood	m³	0.41	21.33	8.79
		Water-based paint	m²	8.4	0.096	0.81
		Steel	t	0.01	85.46	0.75
Wall finishes	Internal plasterboard wall linings to external wall	Plasterboard (10 mm)	m²	3,604	0.207	745
		Water-based paint	m²	3,604	0.096	347.3

Element	Item	Material	Unit	Quantity	Embodied energy coefficient (GJ/unit)	Embodied energy (GJ)
	Wall panelling	MDF/particleboard	m³	6.39	30.35	193.9
		Softwood	m³	1.06	10.93	11.63
		Water-based paint	m²	354.9	0.096	34.21
		Aluminium	t	0.29	252.6	74.27
	Wall tiling	Ceramic tiles	m²	2,790	0.293	818.2
Floor finishes	Insitu terrazzo	32 MPa concrete	m³	120.8	5.81	701.4
	Carpet tiles	Nylon carpet	m²	7,649	0.683	5,226
	Marble	Granite	t	43.23	0.087	3.76
	Ceramic floor tiles	Ceramic tiles	m²	613.2	0.293	179.8
	Vinyl	Vinyl flooring (2 mm)	m²	1,119	0.661	740.2
	Gym floor	Steel	t	1.51	85.46	129.2
		MDF/particleboard	m³	5.44	30.35	165.2
		Nylon carpet	m²	604.8	0.683	413.2
	Access floor	Steel	t	32.69	85.46	2,793
		MDF/particleboard	m³	117.7	30.35	3,572
		Nylon carpet	m²	6,538	0.683	4,467
	Granolithic finish to plant	32 MPa concrete	m³	20.37	5.81	118.3

Element	Item	Material	Unit	Quantity	Embodied energy coefficient (GJ/unit)	Embodied energy (GJ)
	Grid mesh	Aluminium	t	1.35	252.6	340.8
	Mats	Nylon carpet	m²	122.9	0.683	83.94
	Aluminium box skirting	Aluminium	t	0.6	252.6	152.5
	Aluminium skirting	Aluminium	t	2.89	252.6	729.9
Ceiling finishes	Various	Plasterboard (13 mm)	m²	11,628	0.232	2,699
		Water-based paint	m²	11,647	0.096	1,123
		Softwood	m³	5.87	10.93	64.12
		Fibre cement sheet (4.5 mm)	m²	2,356	0.235	554.7
		Vinyl flooring (2 mm)	m²	1,159	0.661	766.6

Total initial embodied energy 194,607

Table B.4 Input-output pathways of *Other construction* sector representing quantified materials for commercial office building

Input-output pathway	Materials	Elements	TER (GJ/A$1000)
Cement, lime and concrete slurry	Concrete	Slab, stairs, external and internal walls, floor finishes	0.79135
Other mining	Sand, screenings, granite	Ground floor slab	0.12908
Ceramic products	Ceramic tiles, clay bricks	Floor finish, internal walls	0.07521
Glass and glass products	Glass	Roof, window and door glazing, screens, balustrades	0.10613
Iron and steel → Fabricated metal products	Steel, stainless steel	Fixings, roof frame, handrails, stair frame, window frames/support, door furniture, wall framing	0.13688
Basic non-ferrous metal and products	Aluminium	Roof canopy and frame, gutters, downpipes, parapet lining, external and internal walls, window and door frames, balustrade	0.00932
Plaster and other concrete products	Plasterboard, FC sheet	External walls, internal walls, ceiling	0.17088
Iron and steel → Structural metal products	Steel	Reinforcement, columns, beam	0.23852
Other wood products	Timber, MDF/particleboard	External and internal doors, internal walls, wall and floor finishes	0.07172
Textile products	Carpet	Carpet tiles, floor covering, mats	0.01504
Sawmill products	Timber	Wall finishes	0.02565
Plastic products	Vinyl flooring, plastic	Internal walls, ceiling, floor, ground floor slab	0.06976
Iron and steel → Sheet metal products	Steel	Floor and roof decking, gutters, downpipes, walkway	0.02042
Paints	Paint	Columns, external and internal walls, external and internal doors, ceilings	0.07235
		Total	1.9323

Table B.5 Input-output pathways of *Residential building* sector representing quantified materials for brick veneer assembly

Element	Input-output pathway (materials covered)	TER (GJ/A$1000)
External cladding	*Ceramic products* (clay bricks)	1.094
	Cement, lime and concrete slurry (mortar)	0.899
Wall ties	*Iron and steel → Fabricated metal products* (wall ties)	0.096
Insulation	*Basic non-ferrous metal and products* (aluminium reflective foil)	0.026
	Other non-metallic mineral products (fibreglass insulation)	0.349
Framing	*Sawmill products* (softwood timber)	0.272
Internal lining	*Plaster and other concrete products* (plasterboard)	0.336
Internal finish	*Paints* (water-based paint)	0.101
	Total	3.173

TER = Total energy requirement. *Note: Iron and steel → Fabricated metal products* represents a demand for goods or services (and thus energy) from the *Iron and steel* sector by the *Fabricated metal products* sector (i.e. the steel used to make wall ties).

Table B.6 Input-output pathways of *Residential building* sector representing quantified materials for timber weatherboard assembly

Element	Input-output pathway (materials covered)	TER (GJ/A$1000)
External cladding and framing	*Sawmill products* (hardwood and softwood timber)	0.272
Insulation	*Basic non-ferrous metal and products* (aluminium reflective foil)	0.026
	Other non-metallic mineral products (fibreglass insulation)	0.349
Internal lining	*Plaster and other concrete products* (plasterboard)	0.336
Internal and external finish	*Paints* (water-based paint)	0.101
	Total	1.084

TER = Total energy requirement.

Table B.7 Initial embodied energy calculation of CRC road construction

Element	Material	Unit	Quantity	Embodied energy coefficient (GJ/unit)	Embodied energy (GJ)
Base	32 MPa concrete	m³	1.33	5.81	7.72
	Steel reinforcement	t	0.133	85.46	11.37
Sub-base	5 MPa concrete	m³	1.38	2.79	3.84
Shoulders	32 MPa concrete	m³	0.76	5.81	4.41
		Total initial embodied energy			27.34

Table B.8 Initial embodied energy calculation of FDA road construction

Element	Material	Unit	Quantity	Embodied energy coefficient (GJ/unit)	Embodied energy (GJ)
Base	Asphalt	m³	2.97	3.08	9.14
Sub-base	Stabilized earth	m³	1.65	2.79	4.60
		Total initial embodied energy			13.74

Table B.9 Initial embodied energy calculation of DSA road construction

Element	Material	Unit	Quantity	Embodied energy coefficient (GJ/unit)	Embodied energy (GJ)
Base	Asphalt	m³	2.22	3.08	6.83
Sub-base	Compacted earth			assumed to be zero	
		Total initial embodied energy			6.83

Table B.10 Initial embodied energy calculation of DSAB road construction

Element	Material	Unit	Quantity	Embodied energy coefficient (GJ/unit)	Embodied energy (GJ)
Base	Asphalt	m³	1.65	3.08	5.08
Sub-base	Stabilized earth	m³	2.2	2.79	6.14
		Total initial embodied energy			11.22

Table B.11 Initial embodied energy calculation of photovoltaic system

Element	Item	Material	Unit	Quantity	Embodied energy coefficient (GJ/unit)	Embodied energy (GJ)
Modules	Aluminium frame	Aluminium	t	0.0286	252.6	7.21
	Foam insulation	Plastic	t	0.0019	156.9	0.295
	Encapsulating glass	Toughened glass (6 mm)	m^2	8.82	3.66	32.24
	Silicon production		m^2	8.82	0.5	4.41
	Silicon purification		m^2	8.82	4.1	36.16
	Crystallization and contouring 1		m^2	8.82	5.7	50.27
	Crystallization and contouring 2		m^2	8.82	2.4	21.17
	Wafering		m^2	8.82	0.25	2.20
	Cell processing		m^2	8.82	0.6	5.29
	Module assembly		m^2	8.82	0.35	3.09
Wiring	Copper wire	Copper	t	0.00492	378.9	1.86
	PVC insulation	Plastic	t	0.000565	156.9	0.089
Inverter			no.	1	0.60	0.60
Displaced roofing	Reflective foil sarking	Aluminium reflective foil	m^2	8.82	0.137	-1.21
	Corrugated steel roofing	Steel decking	m^2	8.82	0.796	-7.02
					Total initial embodied energy	156.67

Table B.12 Input-output pathways representing quantified materials for photovoltaic system

Sector	Element	Input-output pathway (materials covered)	TER (GJ/A$1000)
Other electrical equipment	Modules	Basic non-ferrous metal and products (aluminium frame)	4.1409
		Plastic products (foam insulation)	0.2733
		Glass products (encapsulating glass)	0.0645
		Direct energy (cell production & module assembly)	0.1847
	Wiring	Basic non-ferrous metal and products → Other electrical equipment (copper wire)	0.5073
		Basic chemicals → Other electrical equipment (PVC insulation)	0.2183
		Total	5.3889
Residential building	Displaced roofing	Basic non-ferrous metal and products (reflective foil sarking)	0.0256
		Sheet metal products (corrugated steel roofing)	0.0942
		Total	0.1198

TER = Total energy requirement. *Note: Basic non-ferrous metal and products → Other electrical equipment* represents a demand for goods or services (and thus energy) from the *Basic non-ferrous metal and products* sector by the *Other electrical equipment* sector (i.e. the copper used to make electrical wiring).

Table B.13 Initial embodied energy calculation of wind turbine

Element	Item	Material	Unit	Quantity	Embodied energy coefficient (GJ/unit)	Embodied energy (GJ)
Base	20 MPa reinforced concrete	20 MPa concrete	m³	498.8	4.44	2,212
		Steel reinforcement	t	39.6	85.46	3,384
Tower	Steel, structural	Steel	t	166.7	85.46	14,247
		Water-based paint	m²	8,400	0.096	810
Nacelle	Cables	Aluminium	t	0.72	252.6	183
		Copper	t	0.99	378.9	374
	Cover	Steel	t	9.80	85.46	837
	Frame/bedplate	Steel	t	13.65	85.46	1,167
	Generator	Steel	t	6.00	85.46	512
		Copper	t	1.50	378.9	569
	Brake system	Steel	t	1.07	85.46	92
	Gearbox	Steel	t	24.76	85.46	2,116
		Copper	t	0.25	378.9	96
		Aluminium	t	0.25	252.6	64
	Revolving system	Steel	t	4.06	85.46	347
	Crane	Steel	t	1.07	85.46	92
	Other	Steel	t	3.64	85.46	311

Element	Item	Material	Unit	Quantity	Embodied energy coefficient (GJ/unit)	Embodied energy (GJ)
		Copper	t	1.46	378.9	551
		Aluminium	t	1.46	252.6	368
		Plastic	t	0.69	156.9	109
Rotor	Hub	Steel	t	20.16	85.46	1,723
	Blades	Glass fibre	m³	2.65	432.1	1,144
		Polyester	t	6.62	156.9	1,039
		Epoxy	t	8.43	156.9	1,323
	Bolts	Steel	t	0.77	85.46	66
Miscellaneous	Gear oil	Oil	m³	0.67	34.0	23
	Grease - main-blade, generator bearing	Oil	m³	0.012	34.0	0.41
					Total initial embodied energy	33,757

Table B.14 Input-output pathways representing quantified materials for wind turbine

Sector	Element	Input-output pathway (materials covered)		TER (GJ/A$1000)
Other construction	Base	Cement, lime and concrete slurry (concrete)		0.791
		Iron and steel (steel reinforcement)		0.862
			Total	1.653
Structural metal products	Tower	Iron and steel (steel)		7.367
		Paints (external finish)		0.005
			Total	7.372
Other machinery and equipment	Nacelle	Basic non-ferrous metal and products → Other electrical equipment (aluminium and copper in cables, coolers, transformer and sensors)		0.193
		Iron and steel (steel cover, steel frame/bedplate)		5.263
		Iron and steel → Other electrical equipment (steel and copper in generator, steel in coolers, transformer and sensors)		0.059
		Iron and steel → Other machinery and equipment (steel in brake system, gearbox and revolving system)		0.484
		Basic non-ferrous metal and products → Other machinery and equipment (copper and aluminium in gearbox)		0.034
		Iron and steel → Agricultural, mining and construction machinery, lifting and material handling equipment (steel in crane)		0.278

Sector	Element	Input-output pathway (materials covered)		TER (GJ/A$1000)
		Plastic products → *Other electrical equipment* (plastic in coolers, transformer and sensors)		0.013
			Total	6.324
Plastic products	Rotor - blades	*Basic chemicals* (fibreglass and epoxy)		10.750
			Total	10.750
Fabricated metal products	Rotor - hub	*Iron and steel* (steel bolts and hub)		5.658
			Total	5.658

TER = Total energy requirement. Note: *Iron and steel* → *Other electrical equipment* represents a demand for goods or services (and thus energy) from the *Iron and steel* sector by the *Other electrical equipment* sector (i.e. the steel used to make the turbine generator).

Glossary

Category indicator	reference unit for an impact category
Characterization factor	a factor that is used to convert a life cycle inventory analysis result to the defined reference unit for a particular impact category
Cleaner production	the monitoring and improvement of industrial production processes, including changes in organizational practices and technologies to produce fewer pollutants and less waste and consume fewer non-renewable resources
Delivered energy	the quantity of energy consumed at the point of use, excluding the energy or fuels consumed to produce it (compare to primary energy)
Direct energy requirement	the energy required directly by a process or activity, per unit of product output
Direct requirements coefficient	the value of outputs required directly from one economic sector by another economic sector to produce a certain quantity of output
Downstream truncation	the exclusion of inputs or outputs for a product system lower in the supply chain than a particular product-, process- or activity-related input or output
Eco-label	a labelling system for consumer products used to indicate that they have been designed to minimize negative environmental impacts
Embodied energy	the energy required by all of the activities associated with a production process and the share of energy used in making equipment and other supporting functions (i.e. direct and indirect)
Flyash	a by-product of the coal combustion process
Functional unit	a reference unit for the performance of a product or process
Greenwash	a term used to describe the act of misleading the public regarding the environmental practices of a company or the environmental benefits of a product or service
Hybrid analysis	a life cycle inventory analysis technique for quantifying inputs and outputs of a product, process or activity that combines the use of process- and input-output analysis

Impact category	an approach for organizing environmental impacts into which life cycle inventory analysis results may be grouped or assigned
Indirect energy	the energy required by a process or activity upstream of the main process for the production or supply of supporting goods and services
Input-output analysis	a life cycle inventory analysis technique for quantifying inputs and outputs of a product, process or activity, based on economic input-output tables
Input-output-based hybrid analysis	a life cycle inventory analysis technique for quantifying inputs and outputs of a product, process or activity, based on an input-output analysis framework integrated with data from a process analysis
Inputs	the resources required by a process (e.g. energy or raw materials)
Life cycle	the stages through which something (e.g. a product or a building) passes during its life (usually from raw material acquisition to final disposal)
Life cycle assessment	a tool for measuring the environmental impacts associated with a product, process or activity over its life cycle from raw material acquisition through to production, use and disposal
Life cycle impact assessment	the third phase of a life cycle assessment, which involves the evaluation of the magnitude and significance of potential environmental impacts of a product, process or activity across its life cycle, based on the findings from a life cycle inventory analysis
Life cycle interpretation	the final phase of a life cycle assessment, which involves evaluating the findings from a life cycle inventory analysis and impact assessment so that conclusions and recommendations can be made
Life cycle inventory analysis	the second phase of a life cycle assessment, which involves quantifying the inputs to and outputs from a product across its life cycle
Life cycle study	a study that uses life cycle assessment to quantify the environmental impacts of a product or process across its entire life cycle
Outputs	the waste, emissions, materials and products produced by a process
Pathway	a collection or flow of inputs or outputs to a product system an infinite number of stages upstream of, and

	related to, a particular lower order process in an input-output model
Primary energy	the energy contained within a primary energy source (e.g. coal, oil or natural gas) that has not been processed or converted (compare to delivered energy)
Process analysis	a life cycle inventory analysis technique for quantifying inputs and outputs of a product, process or activity by breaking it down into its constituent parts and tracing the resource requirements and outputs through the supply chain
Process-based hybrid analysis	a life cycle inventory analysis technique for quantifying inputs and outputs of a product, process or activity, based on a process analysis framework with input-output analysis used to fill upstream data gaps
Product system	a collection of processes that perform one or more defined functions
Remainder	those input-output model pathways for which process data has not been substituted, used to fill the data gaps in a process analysis
Sensitivity analysis	a study of the effects of variations in the inputs or parameters of a mathematical model on its output
Sideways truncation	the exclusion of inputs or outputs for a product system at the same level in the supply chain as a particular product-, process- or activity-related input or output
Sustainable	capable of being maintained without long-term negative effects (e.g. on the environment)
Sustainable development	'development that meets the needs of the present without compromising the ability of future generations to meet their own needs' (Brundtland Commission 1987)
System boundary	defines the processes included within a product system
Total energy requirement	the total energy required by a process, per unit of product
Total requirements coefficient	the value of outputs required directly and indirectly from one economic sector by another economic sector to produce a certain quantity of output
Truncation error	the error attributable to the exclusion of processes associated with a product system
Uncertainty analysis	used to assess how the level of uncertainty associated with the particular data or assumptions used within a life cycle assessment affects the reliability of the assessment results

Upstream truncation   the exclusion of inputs or outputs for a product system higher in the supply chain than a particular product-, process- or activity-related input or output

A number of these definitions have been adapted from International Standard 14040 (2006) *Environmental management – life cycle assessment – principles and framework*, Second edition 1 July 2006, Geneva: International Organization for Standardization (ISO).

Notes

1   Figure designed by Philippe Rekacewicz (UNEP/GRID-Arendal) based on information from Okanagan University College, Department of Geography, Canada; University of Oxford, School of Geography; United States Environmental Protection Agency (EPA), Washington; IPCC (1996) *Climate change 1995: the science of climate change. Contribution of working group I to the second assessment report of the Intergovernmental Panel on Climate Change*, UNEP and WMO (World Meteorological Organization), Cambridge, United Kingdom: Cambridge University Press.

2   Whilst the exclusion of a number of life cycle stages in this example is for the purpose of simplifying the assessment, any exclusions in an LCA should typically be well justified and must not significantly change the study findings.

3   Sector names are based on those within the Australian Input-Output Product Classification (IOPC) (Australian Bureau of Statistics (2001b) *Australian National Accounts, Input-Output Tables (Product Details), 1996–97*, ABS Cat. No. 5215.0, Canberra, Australia: Australian Bureau of Statistics.). The categorization of goods and services differs between countries, as do the names of specific sectors.

4   This value represents the *total energy requirement* of the Australian *Residential building* sector (based on Australian Bureau of Statistics (ABS) (2001c) *Energy and Greenhouse Gas Emissions Accounts, Australia: 1992–93 to 1997–98*, ABS Cat. No. 4604.0, Canberra: Australian Bureau of Statistics; and Australian Bureau of Statistics (ABS) (2001a) *Australian National Accounts, Input-Output Tables, 1996–97*, ABS Cat. No. 5209.0, Canberra, Australia: Australian Bureau of Statistics).

References

Alcorn, A. (2003) *Embodied energy and CO₂ coefficients for New Zealand building materials*, Wellington: Centre for Building Performance Research, Victoria University of Wellington.

Alsema, E.A., Frankl, P. and Kato, K. (1998) Energy payback time of photovoltaic energy systems: present status and prospects, *Proceedings: 2nd World Conference on Photovoltaic Solar Energy Conversion*, Vienna, 6-10 July.

Anon. (2009) Plastic wine bottles 'cut emissions', *The Age*, Melbourne, 5 May. Online. Available: <http://news.theage.com.au/breaking-news-national/plastic-wine-bottles-cut-emissions-20090505-atu4.html> (accessed 29 July 2010).

Arbor, A. (1999) Sustainability obstacles, *National Sustainable Buildings Workshop*, October 8-9: 8p.

Asonye, C.C., Okolie, N.P., Okenwa, E.E. and Iwuanyanwu, U.G. (2007) Some physiochemical characteristics and heavy metal profiles of Nigerian rivers, streams and waterways, *African Journal of Biotechnology*, 6(5): 617-24.

ATHENA (2007) *ATHENA® EcoCalculator for Assemblies*, Canada: Athena Sustainable Materials Institute.

ATHENA (2009) *ATHENA® Impact Estimator for Buildings 4.0*, Canada: Athena Sustainable Materials Institute.

Australian Broadcasting Corporation (ABC) (2010) *Green at work: facts and figures*. Online. Available: <http://www.abc.net.au/greenatwork/FactsFigures/#paper> (accessed 10 July 2010).

Australian Bureau of Statistics (ABS) (2001a) *Australian National Accounts, Input-Output Tables, 1996-97*, ABS Cat. No. 5209.0, Canberra, Australia: Australian Bureau of Statistics.

Australian Bureau of Statistics (ABS) (2001b) *Australian National Accounts, Input-Output Tables (Product Details), 1996-97*, ABS Cat. No. 5215.0, Canberra, Australia: Australian Bureau of Statistics.

Australian Bureau of Statistics (ABS) (2001c) *Energy and Greenhouse Gas Emissions Accounts, Australia: 1992-93 to 1997-98*, ABS Cat. No. 4604.0, Canberra: Australian Bureau of Statistics.

Australian Government (1999) *Environment Protection and Biodiversity Conservation Act 1999*, Canberra: Office of Legislative Drafting and Publishing, No.91.

Australian Greenhouse Office (2007) *Energy efficiency in government operations (EEGO) policy*, Canberra: Department of the Environment and Water Resources.

Australian Standard 1085.14 (2003) *Railway track material – Part 14: prestressed concrete sleepers*, 3rd edn, 14 February 2003, Sydney: Standards Australia.

Autodesk (2010) Ecotect analysis sustainable building design software: Autodesk.

Azapagic, A. and Clift, R. (1999) Life cycle assessment and multiobjective optimisation, *Journal of Cleaner Production*, 7(2): 135-43.

BCL (2007) Boustead Model 5.0, United Kingdom: Boustead Consulting Ltd. Available HTTP: <www.boustead-consulting.co.uk>.

Berube, M. and Bisson, S. (1992) *Life cycle studies: a literature review and critical analysis*, Report to Minister of the Environment of Quebec, Quebec: Government of Quebec. January.

Bicknell, K.B., Ball, R.J., Cullen, R. and Bigsby, H.R. (1998) New methodology for the ecological footprint with an application to the New Zealand economy, *Ecological Economics,* 27: 149-60.

Born, P. (1996) Input-output analysis: input of energy, CO_2 and work to produce goods, *Journal of Policy Modeling,* 18: 217-21.

Boustead, I. (1996) Life cycle assessment - an overview, *Proceedings: 1st National Conference on Life Cycle Assessment,* Melbourne, 29 February–1 March.

BP p.l.c. (2010) *BP Statistical Review of World Energy June 2010,* London: BP p.l.c. Available HTTP: <www.bp.com/statisticalreview> (accessed 12 July 2010).

Brown, L.R. (2006) *Plan B 2.0 Rescuing a Planet Under Stress and a Civilization in Trouble,* New York: W.W. Norton & Co.

Brundtland Commission (1987) *Our common future: report of the World Commission on Environment and Development,* Oxford: Oxford University Press.

Building Commission (2008) *Building Code of Australia, 5 Star standard,* Australia: Building Commission.

Bullard, C.W., Penner, P.S. and Pilati, D.A. (1978) Net Energy Analysis: Handbook for Combining Process and Input-Output Analysis, *Resources and Energy,* 1: 267-313.

Bullard, C.W. and Sebald, A.V. (1988) Monte Carlo sensitivity analysis of input-output models, *The Review of Economics and Statistics,* 70(4): 708-12.

Cambridge International Reference on Current Affairs (CIRCA) (2008) *Where we are now: the smartest, clearest guide to the issues that shape the world,* London: Mitchell Beazley. Available: <http://www.holisticpage.com.au/WhereWeAreNow_Circa%7C9781845333447>.

Carnegie Mellon University (2002) *Economic input-output life cycle assessment tool.* Online. Available: <http://www.eiolca.net/> (accessed 5 November 2009).

Carson, R. (1962) *Silent Spring,* Boston: Houghton Mifflin.

Carter, A.J., Peet, N.J. and Baines, J.T. (1981) *Direct and indirect energy requirements of the New Zealand economy - an energy analysis of the 1971-72 inter-industry survey,* Report No. P55, Auckland: New Zealand Energy Research and Development Committee.

Central Intelligence Agency (2009) *The world factbook,* Central Intelligence Agency. Online. Available: <http://purl.access.gpo.gov/GPO/LPS552> (accessed 7 April 2009).

Centre for Design (2001) *Environmental performance data questionnaire for manufacturers of building products,* Melbourne: Royal Melbourne Institute of Technology. Online. Available: <http://buildlca.rmit.edu.au/downloads/EPDS%20 Framework.doc> (accessed 10 May 2010).

City of Melbourne (2006) *Technical research paper 06, energy harvesting systems: economic use and efficiency,* Melbourne. Online. Available: <http://www.melbourne. vic.gov.au> (accessed 10 May 2010).

City of Melbourne (2009) *About CH2.* Online. Available: <http://www.melbourne.vic. gov.au/Environment/CH2/aboutch2/Pages/WaterConservation.aspx> (accessed 12 December 2009).

Cole, R. (1998) Emerging trends in building environmental assessment methods, *Building Research and Information,* 26(1): 3-16.

Cole, R. and Kernan, P.C. (1996) Life cycle energy use in office buildings, *Building and Environment,* 31(4): 307-17.

Cooper, J.S. and Fava, J. (2006) Life cycle assessment practitioner survey: summary of results, *Journal of Industrial Ecology,* 10(4): 12-14.

Cooper, M. (2009) Death toll soared during Victoria's heatwave, *The Age*, Melbourne, 6 April. Online. Available: <http://www.theage.com.au/national/death-toll-soared-during-victorias-heatwave-20090406-9ubd.html> (accessed 28 July 2010).

Crawford, R.H. (2005) Validation of the use of input-output data for embodied energy analysis of the Australian construction industry, *Journal of Construction Research,* 6(1): 71-90.

Crawford, R.H. (2008) Validation of a hybrid life cycle inventory analysis method, *Journal of Environmental Management,* 88(3): 496-506.

Crawley, D. and Aho, I. (1999) Building environmental assessment methods: applications and development trends, *Building Research and Information,* 27(4/5): 300-8.

Crowther, P. (1999) Design for disassembly to recover embodied energy, *Proceedings of the 16th International Conference on Passive and Low Energy Architecture,* Melbourne, September.

Curran, M.A. (1993) Broad based environmental life cycle assessment, *Environmental Science and Technology,* 27(3): 430-6.

Daly, H.E. (1990) Toward some operational principles of sustainable development, *Ecological Economics,* 2(1): 1-6.

Deni Greene Consulting Services (1992) *Life cycle analysis: a view of the environmental impact of consumer products using clothes washing machines as an example,* Marrickville: Australian Consumers' Association.

Department of Climate Change (2008) *National greenhouse and energy reporting (measurement) determination 2008,* Canberra: Commonwealth of Australia.

Department of Environment and Heritage (DEH) (2006) *Waste and recycling in Australia,* Report no. 4, Melbourne: Hyder Consulting.

Department of Infrastructure, Energy and Resources (DIER) (2009) *Infrastructure and resource information service (IRIS) Tasmania: energy supply.* Online. Available: <http://www.iris.tas.gov.au/infrastructure/energy/supply> (accessed 12 February 2010).

Department of Sustainability and Environment (DSE) (2010) *Our water our future: business and industry.* Online. Available: <http://www.ourwater.vic.gov.au/saving/industry> (accessed 28 June 2010).

Department of the Environment, Water, Heritage and the Arts (DEWHA) (2009) *National pollutant inventory 2006-07 summary report,* Canberra: Commonwealth of Australia.

Department of the Environment, Water, Heritage and the Arts (DEWHA) (2010) *National pollutant inventory 2008-09 emission data,* Canberra: Commonwealth of Australia. Online. Available HTTP: <www.npi.gov.au> (accessed 1 July 2010).

Dixon, R. (1996) Inter-industry transactions and input-output analysis, *The Australian Economic Review,* 3(115): 327-36.

Duchin, F. (1992) Industrial input-output analysis: implications for industrial ecology, *Proceedings of the National Academy of Science of the USA*: 851-5.

Ecquate (2009) LCADesign software, Thirroul, Australia: Ecquate Pty Ltd.

Elliot, M. and Thomas, I. (2009) *Environmental Impact Assessment in Australia: theory and practice,* 5th edn, Sydney: The Federation Press.

Energy Information Administration (EIA) (2001) *Manufacturing energy consumption survey 1998,* Washington: Office of Energy Markets and End Use, Energy Consumption Division.

Energy Information Administration (EIA) (2008a) *International energy annual 2006: world consumption of primary energy by energy type and selected country groups,*

Washington: Energy Information Administration (EIA). Online. Available: <http://www.eia.doe.gov/iea> (accessed 17 June 2010).

Energy Information Administration (EIA) (2008b) *International energy statistics: total carbon dioxide emissions from the consumption of energy*, Washington: Energy Information Administration (EIA). Online. Available: <http://tonto.eia.doe.gov/cfapps/ipdbproject/IEDIndex3.cfm> (accessed 17 June 2010).

European Commission (2010) *Life cycle thinking and assessment*, Institute for the Environment and Sustainability. Online. Available: <http://lca.jrc.ec.europa.eu/lcainfohub/toolList.vm> (accessed 16 June 2010).

European Committee for Standardization (CEN) (2010) *Sustainability of construction works - assessment of buildings*, CEN/prEN 15643: under approval.

European Environment Agency (2010) *Waterbase - emissions to water*, The European Topic Centre on Water. Online. Available: <http://www.eea.europa.eu/data-and-maps/data/waterbase-emissions> (accessed 1 July 2010).

European Union (1985) Council Directive 85/337/EEC on the Assessment of the Effects of Certain Public and Private Projects on the Environment, 27 June 1985.

Evans, D. and Ross, S. (1998) The role of life cycle assessment in Australia, *Australian Journal of Environmental Management*, 5(3): 137-45.

Fay, R., Treloar, G. and Iyer-Raniga, U. (2000) Life cycle energy analysis of buildings: a case study, *Building Research and Information*, 28(1): 31-41.

Finnveden, G., Hauschild, M.Z., Ekvall, T., Guinee, J.B., Heijungs, R., Hellweg, S., Koehler, A., Pennington, D. and Suh, S. (2009) Recent developments in life cycle assessment, *Journal of Environmental Management*, 91: 1-21.

Forest & Wood Products Australia (FWPA) (2007) *Timber service life design guide*, PN07.1052, Melbourne: FWPA.

Fuller, R.J., Crawford, R.H. and Leonard, D. (2009) What is wrong with a big house?, *Performative Ecologies in the Built Environment: Sustainable Research across Disciplines: Proceedings of the 43rd Annual Conference of the Australian and New Zealand Architectural Science Association ANZAScA*, Launceston, November.

Garden, G.K. (1980) Design determines durability, *Proceedings of the First International Conference on Durability of Building Materials and Components*, Ottawa, Canada, 21-3 August 1978: 31-7.

Gari, L. (2002) Arabic treatises on environmental pollution up to the end of the thirteenth century, *Environment and History*, 8(4): 475-88.

Goodman, G.T. (1974) How do chemical substances affect the environment?, *Proceedings of the Royal Society of London, Series B, Biological Sciences*, 185(1079): 127-48.

Gorbachev, Mikhail (n.d.) Better World. Online. Available: <http://www.betterworldheroes.com/pages-g/gorbachev-quotes.htm> (accessed 15 September 2010).

Grant, T. (2002) *Australian material inventory database of life cycle assessment values for materials*, Melbourne: RMIT.

Hansen, J., Sato, M., Kharecha, P., Beerling, D., Berner, R., Masson-Delmotte, V., Pagani, M., Raymo, M., Royer, D.L. and Zachos, J.C. (2008) Target atmospheric CO2: where should humanity aim?, *Open Atmospheric Science Journal*, 2: 217-31.

Heijungs, R., Guinee, J.B., Huppes, G., Lankreijer, R.M., Udo de Haes, H.A., Wegener Sleeswijk, A., Ansems, A.M.M., Eggels, P.G., van Duin, R. and Goede, H.P. (1992) *Environmental life cycle assessment of products*, Leiden: CML Centre for Environmental Studies, Leiden University.

Hendrickson, C.T., Horvath, A., Joshi, S. and Lave, L.B. (1998) Economic input-output models for environmental life cycle assessment, *Environmental Science and Technology,* 32(7): 184A-91A.

Hill, R.K. (1978) Gross energy requirements of building materials, *Proceedings: Conference on Energy Conservation in the Built Environment,* Department of Environment, Housing and Community Development Conference, Sydney, 15-16 March: 179-90.

Holness, G.V.R. (2008) Improving energy efficiency in existing buildings, *ASHRAE Journal,* American Society of Heating, Refrigerating, and Air-Conditioning Engineers, Inc., January.

Hunt, R.G., Franklin, W.E., Welch, R.O., Cross, J.A. and Woodall, A.E. (1974) *Resource and environmental profile analysis of nine beverage container alternatives,* EPA/530/ SW-91c, Washington: United States Environmental Protection Agency, Office of Solid Waste Management Programs.

International Energy Agency (2004) *Annex 31: Energy-related environmental impact of buildings, report 4: LCA methods for buildings,* Ontario: Canada Mortgage and Housing Corporation. Available: <http://www.iisbe.org/annex31/index.html> (accessed 10 May 2010).

International Standard 14025 (2006) *Environmental labels and declarations - type III environmental declarations - principles and procedures,* 1st edn, 1 July 2006, Geneva: International Organization for Standardization (ISO).

International Standard 14040 (1997) *Environmental management - life cycle assessment - principles and framework,* 1st edn, 15 June 1997, Geneva: International Organization for Standardization (ISO).

International Standard 14040 (2006) *Environmental management - life cycle assessment - principles and framework,* 2nd edn, 1 July 2006, Geneva: International Organization for Standardization (ISO).

International Standard 14041 (1998) *Environmental management - life cycle assessment - goal and scope definition and inventory analysis,* 1st edn, 1 October 1998, Geneva: International Organization for Standardization (ISO).

International Standard 14042 (2000) *Environmental management - life cycle assessment - life cycle impact assessment,* 1st edn, 1 March 2000, Geneva: International Organization for Standardization (ISO).

International Standard 14043 (2000) *Environmental management - life cycle assessment - life cycle interpretation,* 1st edn, 1 March 2000, Geneva: International Organization for Standardization (ISO).

International Standard 14044 (2006) *Environmental management - life cycle assessment - requirements and guidelines,* 1st edn, 1 July 2006, Geneva: International Organization for Standardization (ISO).

International Standard 15686-5 (2008) *Buildings and constructed assets: service-life planning, part 5: life cycle costing,* Geneva: International Organization for Standardization (ISO).

IPCC (2007a) *Climate change 2007: impacts, adaptation and vulnerability. Contribution of working group II to the fourth assessment report of the Intergovernmental Panel on Climate Change,* M.L. Parry, O.F. Canziani, J.P. Palutikof, P.J. van der Linden, C.E. Hanson (eds), Cambridge: Cambridge University Press.

IPCC (2007b) *Climate change 2007: the physical science basis. Contribution of working group I to the fourth assessment report of the Intergovernmental Panel on Climate Change,* S. Solomon, D. Qin, M. Manning, Z. Chen, M. Marquis, K.B. Averyt, M. Tignor, H.L. Miller (eds), Cambridge and New York: Cambridge University Press.

IPCC (2007c) *Climate change 2007: mitigation. Contribution of working group III to the fourth assessment report of the Intergovernmental Panel on Climate Change*, B. Metz, O.R. Davidson, P.R. Bosch, R. Dave, L.A. Meyer (eds), Cambridge and New York: Cambridge University Press.

Isard, W., Bassett, K., Choguill, C., Furtado, J., Izumita, R., Kissin, J., Romanoff, E., Seyfarth, R. and Tatlock, R. (1968) On the linkage of socio-economic and ecologic systems, *Papers and Proceedings of the Regional Science Association*, 21(1): 79-99.

Isard, W. and Romanoff, E. (1967) *Water use and water pollution coefficients: preliminary report*, Technical Paper No. 6, Cambridge, MA: Regional Science Research Institute.

Jackson, B., Lewis, M. and Stock, A.L. (2010) *4th annual green building survey*, Wahington, DC: Allen Matkins, Constructive Technologies Group (CTG) and Green Building Insider.

Jensen, A.A., Elkington, J., Christiansen, K., Hoffmann, L., Moller, B.T., Schmidt, A. and Dijk, F.V. (1997) *Life cycle assessment (LCA): a guide to approaches, experiences and information sources*, Søborg, Denmark: TEKNIK Energy & Environment.

Jones, J.A.T., Bowman, B. and Lefrank, P.A. (1998) Electric furnace steelmaking, *The making, shaping and treating of steel*, R.J. Fruehan, Pittsburgh: The AISE Steel Foundation: 525-660.

Jönsson, A. (2000) Tools and methods for environmental assessment of building products - methodological analysis of six selected approaches, *Building and Environment*, 35(3): 223-38.

Joshi, S. (1999) Product environmental life cycle assessment using input-output techniques, *Journal of Industrial Ecology*, 3(2/3): 95-120.

Junnila, S. and Horvath, A. (2003) Life cycle environmental effects of an office building, *Journal of Infrastructure Systems*, 9(4): 157-66.

Kien, H.L. and Ofori, G. (2002) Minimizing environmental impacts of building materials in Singapore: role of architects, *International Journal of Environmental Technology and Management, Special Issue: sustainable built environments*, 2(1-3).

Kohn, R.E. (1972) Input-output analysis and air pollution control, *Proceedings: Economic Analysis of Environmental Problems*, Chicago, Illinois.

Langston, C. (2005) *Life-cost Approach to Building Evaluation*, Sydney: Elsevier/UNSW Press.

Langston, C., Wong, F., Hui, E. and Shen, L.Y. (2008) Strategic assessment of building adaptive reuse opportunities in Hong Kong, *Building and Environment*, 43(10): 1709-18.

Lave, L.B., Cobas-Flores, E., Hendrickson, C.T. and McMichael, F. (1995) Life cycle assessment: using input-output analysis to estimate economy-wide discharges, *Environmental Science and Technology*, 29(9): 420A-6A.

Lenzen, M. (2001) Errors in conventional and input-output-based life cycle inventories, *Journal of Industrial Ecology*, 4(4): 127-48.

Lenzen, M. and Dey, C.J. (2000) Truncation error in embodied energy analyses of basic iron and steel products, *Energy*, 25: 577-85.

Lenzen, M. and Lundie, S. (2002) *Input-output model of the Australian economy based on published 1996-97 Australian input-output data*, Sydney: University of Sydney.

Leontief, W. and Ford, D. (1970) Environmental repercussions and the economic structure: an input-output approach, *Review of Economics and Statistics*, 52: 262-71.

Lewis, H. and Demmers, M. (1996) Life cycle assessment and environmental management, *Australian Journal of Environmental Management*, 3(2): 110-23.

Meadows, D.H., Meadows, D.L., Randers, J. and Behrens III, W.W. (1972) *The limits to growth: a report for the Club of Rome's project on the predicament of mankind*, New York: Universe Books.

Meinshausen, M. (2006) What does a 2°C target mean for greenhouse gas concentrations? A brief analysis based on multi-gas emission pathways and several climate sensitivity uncertainty estimates, *Avoiding dangerous climate change*, H.J. Schellnhuber, W. Cramer, N. Nakicenovic, T. Wigley, G. Yohe (eds), Cambridge: Cambridge University Press, pp. 253-80.

Metricon (2010) *Floor plan of Bel-Air house design*, Metricon Pty Ltd. Online. Available HTTP: <www.metricon.com.au> (accessed 18 March 2010).

Miller, M.A.L. (1995) *The Third World in Global Environmental Politics*, Buckingham: Open University Press.

Miller, R.E. and Blair, P.D. (1985) *Input-Output Analysis - Foundations and Extensions*, Upper Saddle River, NJ: Prentice Hall.

Moskowitz, P.D. and Rowe, M.D. (1985) A comparison of input-output and process analysis, in P.F. Ricci and M.D. Rowe (eds), *Health and Environmental Risk Assessment*, New York: Pergamon Press, pp. 281-93.

Myers, G., Reed, R. and Robinson, J. (2008) Investor perception of the business case for sustainable office buildings: evidence from New Zealand, *Proceedings: 14th Annual Pacific Rim Real Estate Society Conference: Investing in Sustainable Real Estate Environment*, Kuala Lumpur, Malaysia, 20-23 January.

National Institute of Standards and Technology (NIST) (2007) *Building for environmental and economic sustainability version 4.0*, Washington: United States National Institute of Standards and Technology.

National Wind Coordinating Committee (NWCC) (2001) *Avian collisions with wind turbines: a summary of existing studies and comparisons to other sources of avian collision mortality in the United States*, Washington: National Wind Coordinating Committee, Report prepared by WEST Inc. August.

Norris, G. (2001) Empirically derived distributions of life cycle emissions, in B.P. Weidema and A.M. Nielsen (eds), *Input/Output analysis: shortcuts to life cycle data?*, Copenhagen, Denmark: Environmental Project No. 581, Ministry for Environment and Energy, pp. 52-6.

OECD (2003) *Environmentally Sustainable Buildings - Challenges and Policies*, Paris: Organisation for Economic Co-Operation and Development.

OECD (2005) *Environment at a glance: OECD environmental indicators*, Paris: Organisation for Economic Co-operation and Development.

OECD (2008) *OECD environmental outlook to 2030*, Paris: Organisation for Economic Co-operation and Development.

Ofori, G. (1998) Sustainable construction: principles and a framework for attainment - comment, *Construction Management and Economics*, 16(2): 141-5.

Ok Tedi Mining Limited (OTML) (2003) *Impacts of mining*. Online. Available: <http://www.oktedi.com/community-and-environment/the-environment/impacts-of-mining> (accessed 24 June 2010).

Packaging Digest (2008) *Survey says 'green' is growing*, Survey by Packaging Digest and the Sustainable Packaging Coalition. Online. Available HTTP: <www.packagingdigest.com/article/CA6610094.html> (accessed 25 July 2010).

Pedersen, B. and Christiansen, K. (1992) A meta-review on product life assessment, *Product Life Cycle Assessment*, Copenhagen and Stockholm: Nordic Council of Ministers and The Nordic Council.

PE International (2010) GaBi software 4.3, Germany: PE International. Available HTTP: <www.gabi-software.com>.

Petit, J.R., Jouzel, J., Raynaud, D., Barkov, N.I., Barnola, J.-M., Basile, I., Bender, M., Chappellaz, J., Davis, M., Delaygue, G., Delmotte, M., Kotlyakov, V.M., Legrand, M., Lipenkov, V.Y., Lorius, C., Pepin, L., Ritz, C., Saltzman, E. and Stievenard, M. (1999) Climate and atmospheric history of the past 420,000 years from the Vostok ice core, Antarctica, *Nature*, 399(6735): 429-36.

PRé Consultants (2010) SimaPro 7.2, The Netherlands: PRé Consultants. Available: <www.pre.nl>.

Productivity Commission (2006) *Waste management*, Report no. 38, Canberra: Commonwealth of Australia.

Proops, J.L.R. (1977) Input-output analysis and energy intensities: a comparison of methodologies, *Applied Mathematical Modelling*, 1(March): 181-6.

Pullen, S. (1995) *Embodied energy of building materials in houses*, Master of Building Science Thesis, University of Adelaide, Adelaide: 184p.

Pullen, S. and Perkins, A. (1995) Energy use in the urban environment and its greenhouse gas implications, *Transactions of the International Symposium on Energy, Environment and Economics*, University of Melbourne, 20-24 November: 383-9.

Rebitzer, G. and Schafer, J.H. (2009) The remaining challenge - mainstreaming the use of LCA, *The International Journal of Life Cycle Assessment*, 14(May): 101-2.

Rockström, J., Steffen, W., Noone, K., Persson, Å., Chapin, F.S., Lambin, E.F., Lenton, T.M., Scheffer, M., Folke, C., Schellnhuber, H.J., Nykvist, B., de Wit, C.A., Hughes, T., van der Leeuw, S., Rodhe, H., Sörlin, S., Snyder, P.K., Costanza, R., Svedin, U., Falkenmark, M., Karlberg, L., Corell, R.W., Fabry, V.J., Hansen, J., Walker, B., Liverman, D., Richardson, K., Crutzen, P. and Foley, J.A. (2009) A Safe Operating Space For Humanity, *Nature*, 461(24 September): 472-5.

Sandler, K. (2003) Analyzing what's recyclable in C&D debris, *BioCycle*, 44(11): 51-4.

Scientific Applications International Corporation (SAIC) (2006) *Life cycle assessment: principles and practice*, Report no. EPA/600/R-06/060, Cincinnati: US Environmental Protection Agency.

Sebald, A.V. (1974) *An analysis of the sensitivity of large scale input-output models to parametric uncertainties*, CAC Document No. 122, Illinois: Centre for Advanced Computation, University of Illinois at Urbana-Champaign.

Smith, V.H., Tilman, G.D. and Nekola, J.C. (1999) Eutrophication: impacts of excess nutrient inputs on freshwater, marine, and terrestrial ecosystems, *Environmental Pollution*, 100(1-3): 179-96.

Society of Environmental Toxicology and Chemistry (SETAC) (1993) *Guidelines for life cycle assessment: a 'code of practice'*, F. Consoli, D. Allen, I. Boustead, N. de Oude, J. Fava, R. Franklin, A.A. Jensen, R. Parrish, R. Perriman, D. Postlethwaite, B. Quay, J. Séguin, B. Vigon (eds), Report from a SETAC Workshop in Sesimbra, Brüssels.

Sonneveld, K. (2000) Drivers and barriers for LCA penetration in Australia, *Proceedings: 2nd National Conference on LCA*, Melbourne.

Sorrell, S., O'Malley, E., Schleich, J. and Scott, S. (2004) *The Economics of Energy Efficiency*, Cheltenham: Edward Elgar.

Strømman, A.H., Peters, G.P. and Hertwich, E.G. (2009) Approaches to correct for double counting in tiered hybrid life cycle inventories, *Journal of Cleaner Production*, 17(2): 248–54.

Suh, S. (2002) The hybrid approach merging IO and process LCA, *Proceedings: 16th Discussion Forum on Life Cycle Assessment*, Lausanne, 9 April.

Suh, S. and Huppes, G. (2002) Missing inventory estimation tool using input-output analysis, *International Journal of Life Cycle Assessment,* 7(3): 134-40.

Suh, S., Lenzen, M., Treloar, G.J., Hondo, H., Horvath, A., Huppes, G., Jolliet, O., Klann, U., Krewitt, W., Moriguchi, Y., Munksgaard, J. and Norris, G. (2004) System boundary selection in life cycle inventories, *Environmental Science and Technology,* 38(3): 657-64.

Sustainability Victoria (2008) *Victorian recycling industries annual survey 2006–2007,* Melbourne: Sustainability Victoria.

Suzuki, M. and Oka, T. (1998) Estimation of life cycle energy consumption and CO_2 emission of office buildings in Japan, *Energy and Buildings,* 28: 33-41.

Tan, A.T.K., Ofori, G. and Briffett, C. (1999) ISO 14000: its relevance to the construction industry of Singapore and its potential as the next industry milestone, *Construction Management and Economics,* 17(4): 449-61.

Tans, P. (2010) *Trends in atmospheric carbon dioxide - global,* Boulder, CO: National Oceanic & Atmospheric Administration, Earth System Research Laboratory, Global Monitoring Division. Online. Available: <www.esrl.noaa.gov/gmd/ccgg/trends> (accessed 4 July 2010).

The EarthWorks Group (1990) *The Recycler's Handbook: simple things you can do,* Berkeley, CA: EarthWorks Press.

Tillman, A., Ekvall, T., Baumann, H. and Rydberg, T. (1994) Choice of system boundaries in life cycle assessment, *Journal of Cleaner Production,* 2(1): 21-9.

Todd, J.A. and Curran, M.A. (1999) *Streamlined life cycle assessment: a final report from the SETAC North America streamlined LCA workgroup,* Pensacola: FL: Society of Environmental Toxicology and Chemistry (SETAC).

Treloar, G.J. (1997) Extracting embodied energy paths from input-output tables: towards an input-output-based hybrid energy analysis method, *Economic Systems Research,* 9(4): 375-91.

Treloar, G.J. (1998) *A comprehensive embodied energy analysis framework,* Ph.D. Thesis, Deakin University, Geelong.

Treloar, G.J. (2000) Streamlined life cycle assessment of domestic structural wall members, *Journal of Construction Research,* 1: 69-76.

Treloar, G.J. (2007) Environmental assessment using both financial and physical quantities, *Proceedings of the 41st Annual Conference of the Architectural Science Association ANZAScA,* Geelong, November: 247-55.

Treloar, G.J. and Crawford, R.H. (2010) *Database of embodied energy and water values for materials,* Melbourne: The University of Melbourne.

Treloar, G.J., Fay, R., Love, P.E.D. and Iyer-Raniga, U. (2000a) Analysing the life cycle energy of an Australian residential building and its householders, *Building Research and Information,* 28(3): 184-95.

Treloar, G.J., Gupta, H., Love, P.E.D. and Nguyen, B. (2003) An analysis of factors influencing waste minimisation and use of recycled materials for the construction of residential buildings, *Management of Environmental Quality,* 14(1): 134-45.

Treloar, G.J., Love, P.E.D., Faniran, O.O. and Iyer-Raniga, U. (2000b) A hybrid life cycle assessment method for construction, *Construction Management and Economics,* 18: 5-9.

Treloar, G.J., Love, P.E.D. and Holt, G.D. (2001) Using national input-output data for embodied energy analysis of individual residential buildings, *Construction Management and Economics,* 19: 49-61.

UNEP/GRID-Arendal (2002) *Greenhouse effect*, UNEP/GRID-Arendal Maps and Graphics Library, Designer: Philippe Rekacewicz. Online. Available: <http://maps.grida.no/go/graphic/greenhouse-effect> (accessed 24 July 2010).

University of Bath (2008) *Inventory of carbon and energy (ICE) version 1.6a*, Bath, United Kingdom. Online. Available: <http://www.bath.ac.uk/mech-eng/sert/embodied> (accessed 10 May 2010).

US Census Bureau (2010a) *International data base, world population trends: total midyear population for the world 1950-2050*. Online. Available: <http://www.census.gov/ipc/www/idb/worldpop.php> (accessed 1 July 2010).

US Census Bureau (2010b) *International data base, world population trends: historical estimates of world population*. Online. Available: <http://www.census.gov/ipc/www/worldhis.html> (accessed 1 July 2010).

US Geological Survey (2005) *Water use in the United States*, US Geological Survey. Online. Available: <http://water.usgs.gov/watuse> (accessed 5 June 2010).

US Geological Survey (2009) *Mineral commodity summaries*, US Geological Survey. Online. Available: <http://minerals.usgs.gov/minerals/pubs/mcs/> (accessed 28 July 2010).

US Government (1969) *National Environmental Policy Act of 1969*, 42 USC, 91-190.

Verbeek, I. and Wibberley, L. (1996) Data collection, *1st National Conference on Life Cycle Assessment*, Melbourne, 29 February–1 March.

Vestas Wind Systems (2006) *Life cycle assessment of offshore and onshore sited wind power plants based on Vestas V90-3.0 MW turbines*, Denmark: Vestas Wind Systems.

Voorspools, K.R., Brouwers, E.A. and D'haeseleer, W.D. (2000) Energy content and indirect greenhouse gas emissions embedded in 'emission-free' power plants: results for the low countries, *Applied Energy*, 67: 307-30.

WasteCap (2010) *Office waste reduction briefing paper: downsizing paperwork; increasing productivity: fact sheet*, Wisconsin. Online. Available: <http://www.wastecapwi.org/documents/officewaste.pdf> (accessed 28 July 2010).

Weidema, B. (1997) *Environmental Assessment of Products: A Textbook on Life Cycle Assessment*, Finland: The Finnish Association of Graduate Engineers TEK.

Weidema, B. and Wesnaes, M.S. (1996) Data quality management for life cycle inventories - an example of using data quality indicators, *Journal of Cleaner Production*, 4(3-4): 167-74.

World Business Council for Sustainable Development (WBCSD) (2007) *Energy efficiency in buildings: business realities and opportunities, summary report*, Washington: World Business Council for Sustainable Development.

World Health Organization (WHO) (2002) *The world health report 2002 - reducing risks, promoting healthy life*, Geneva: World Health Organization.

Worth, D.T. (1993) Embodied energy analysis of buildings part 1: determining the energy content of building materials, *Exedra - Architecture, Art and Design*, 4(1): 6-10.

Zapata, P. and Gambatese, J.A. (2005) Energy consumption of asphalt and reinforced concrete pavement materials and construction, *Journal of Infrastructure Systems*, 11(1): 9-20.

Index

Milton Keynes UK
Ingram Content Group UK Ltd.
UKHW040108071024
449327UK00019B/917